WIND ENERGY TECHNOLOGY

UNESCO ENERGY ENGINEERING SERIES

ENERGY ENGINEERING LEARNING PACKAGE

Organized by UNESCO, this innovative distance learning package has been established to train engineers to meet the challenges of today and tomorrow in this exciting field of energy engineering. It has been developed by an international team of distinguished academics coordinated by Dr Boris Berkovski. This modular course is aimed at those with a particular interest in renewable energy, and will appeal to advanced undergraduate and postgraduate students, as well as practising power engineers in industry.

Solar Electricity
Edited by Tomas Markvart

Magnetohydrodynamic Electrical Power Generation
Hugo Messerle

Geothermal Energy
Edited by Mary H. Dickson and Mario Fanelli

Energy Planning and Policy
Maxime Kleinpeter

Ocean Thermal Energy Conversion
Patrick Takahashi and Andrew Trenka

Industrial Energy Conservation
Charles M. Gottschalk

Biomass Conversion and Technology
Charles Y. Wereko-Brobby and Essel B. Hagan

Mini Hydropower
Tong Jiandong et al.

Wind Energy Technology
John F. Walker and Nicholas Jenkins

WIND ENERGY TECHNOLOGY

John F. Walker
Wood Gen, Waltham Chase

Nicholas Jenkins
UMIST, Manchester, UK

JOHN WILEY & SONS
Chichester • New York • Weinheim • Brisbane • Toronto • Singapore

John Wiley & Sons
ster,
JD, England

79777
243 779777

quiries): cs- books @ wiley.co.uk

co.uk

om

Any opinions expressed in this text reflect exclusively those of the authors and are not necessarily those of UNESCO or any other affiliated organization.

UNESCO gratefully acknowledges the financial contributions and other forms of co-operation received from the International Technological University (ITU) London and Paris, over the period 1991–1994 towards several of the texts in this Series. Without the consent and goodwill of the ITU Board of Trustees, and especially Dr S.M.A. Shahrestani, one of its founding members, some of the texts could not have been produced.

Other Wiley Editorial Offices

John Wiley & Sons, Inc., 605 Third Avenue,
New York, NY 10158–0012, USA

VCH Verlagsgesellschaft mbH, Pappelallee 3,
D-69469 Weinheim, Germany

Jacaranda Wiley Ltd, 33 Park Road, Milton,
Queensland 4064, Australia

John Wiley & Sons (Asia) Pte Ltd, 2 Clementi Loop #02-01,
Jin Xing Distripark, Singapore 0512

John Wiley & Sons (Canada) Ltd, 22 Worcester Road,
Rexdale, Ontario M9W 1L1, Canada

Library of Congress Cataloging-in-Publication Data

Walker, J. F.
Wind energy / J. F. Walker, N. Jenkins.
p. cm. – (UNESCO energy engineering series)
Includes bibliographical references and index.
ISBN 0–471–96044–6
1. Wind Power. I. Jenkins, N. II. Title. III. Series.
TJ820.W35 1997
621.4′5 — dc21 96–40504
CIP

British Library Cataloguing in Publication Data

A catalogue record for this book is available from the British Library

ISBN 0 471 96044 6

Typeset in 10/12 pt Times by Pure Tech India Ltd, Pondicherry
Printed and bound in Great Britain by Bookcraft (Bath) Ltd
This book is printed on acid-free paper responsibly manufactured from sustainable forestation, for which at least two trees are planted for each one used for paper production.

Contents

Foreword

Education applied to a whole complex of interlocking problems is the master key that can open the way to sustainable development.

A major obstacle to the development of renewable energies is the paucity of relevant information available to engineers and technicians (who too often lack the necessary know-how and skills) as well as to decision-makers and users. One of UNESCO's priorities in this domain is therefore to promote training and information aimed at sensitizing specialists and the general public to the possible uses of renewable energy sources, with particular regard to environmental concerns and the requirements of sustainable development.

The present textbook on *Wind Energy Technology* in UNESCO's Energy Engineering Series is designed not only to provide instruction for a new generation of engineers but also to foster the kind of environmental management so urgently needed throughout the world. The series as a whole is a venture involving many institutions working together – on the basis of pre-defined standards – to promote awareness of the environmental, cultural, economic and social dimensions of renewable energy issues, as well as knowledge of those technical aspects that form the core of the UNESCO Learning Package Programme.

The highly successful World Solar Summit, held in Harare, Zimbabwe, in September 1996, endorsed five activities of universal scope and value as part of the World Solar Programme 1996–2005. One of these is the *Global Renewable Energy Education and Training Programme*, of which this Learning Package constitutes a first and significant element. The UNESCO Energy Engineering Series should make a useful contribution to the World Solar Programme 1996–2005, and UNESCO is pleased to be a partner in its development.

Federico Mayor
Director-General, UNESCO

Acknowledgements

The initial preparation of the manuscript was based on the UNESCO Postgraduate Course in Energy Engineering: Wind Energy 1991 (COWIconsult, Consulting Engineers and Planners, Denmark).

The authors would like to acknowledge the substantial contribution of Dr David Hoadley, formerly of the University of Southampton, in particular to the chapter on aerodynamics and to the preparation of many of the self-assessment questions.

We would also like to thank Mark Hancock of Taywood Aerolaminates for his contribution to the discussion of wind turbine blades.

Preface

Windmills have been in existence for at least three thousand years, mainly for grinding grain or pumping water. The earliest machines were various forms of the vertical-axis type; the first horizontal-axis windmills appeared in the twelfth century. From the nineteenth century new designs of windpump evolved for agricultural use in a form which is still common today. But following the advent of cheap fossil fuels, to provide electrical or mechanical power, the use of windmills declined generally, with a few notable exceptions.

The use of windmills to generate electricity is a twentieth-century development, initially at low power levels for uses such as battery charging. In the early 1970s, the first oil-price shock stimulated interest in alternative energy sources and research and development into modern wind turbine technology received a substantial boost in several countries. Increasing concern for the environment provided a further impetus for cleaner sources of energy, including wind energy. These two factors promoted interest in the large-scale generation of electricity by wind power, until by 1996 some 6000 MW of wind plant were either installed or were being planned in countries throughout the world.

The variability of the wind is an obvious fact, but understanding the wind and its conversion into useful energy are only part of the story. The planning of a wind project requires consideration of economics and environmental issues. A wind farm will require connection into the electricity network. Effective management is essential for the successful realisation of wind projects.

This book introduces students to modern wind turbine technology; it covers mainly horizontal-axis machines, since their use is now almost universal, although brief reference is made to vertical-axis machines. The book is divided into two parts. Part A deals with wind turbine theory and its application. It begins by outlining the characteristics of the wind as an energy resource, including treatment of its variability, and the principles of its conversion into useful energy. Wind turbine aerodynamics are covered in sufficient detail for the reader to gain an understanding of the underlying theoretical principles and how they are applied in practice. It also describes the components and operational characteristics of typical modern wind turbines.

Part B is concerned with project assessment and engineering. It begins with an introduction to economic assessment and to planning, environmental and social issues. Some typical applications are described, including large-scale electricity generation, wind–diesel systems and water pumping. Finally, there are chapters on project engineering and international and national wind energy programmes.

After working through the book, including the extensive self-assessment questions, the student will have a broad appreciation of modern wind energy technology and how it is applied in practice.

Notation and units

RELEVANT SI SYMBOLS*

K	degree Kelvin (temperature)
kg	kilogramme (mass)
m	metre (length)
N	newton (force)
s	second (time)
W	watt, joule per second (power)

SI PREFIXES

m	milli	10^{-3}
k	kilo	10^{3}
M	mega	10^{6}
G	giga	10^{9}
T	tera	10^{12}

* Specific notation and units are listed at the beginning of each chapter.

PART A

Wind Turbine Theory and Practice

It is essential for the student to gain a sound, basic understanding of the wind and how its energy can be converted. The variability of the wind is a matter of everyday experience. It is obvious there will be times when there is no wind, so wind turbines will not turn. At other times the wind may be strong and a wind turbine will be able to operate at full output. In some places the average wind speeds will be high, such as in parts of southern Spain or on the western coasts of the British Isles.

Wind speed and direction vary in the short and the long term. Gusts occur over a period of seconds; average wind speeds may change with time of day or the season of the year. Some years may be exceptionally windy, others calmer than average. Wind speed is affected by the terrain and by height above ground. The power of a wind turbine varies approximately as the cube of the wind speed over much of its operating range, so doubling the wind speed increases power by a factor of 8. Overestimates of wind speed will result in less wind energy being available than expected, which may be disastrous for the economics of a project. Underestimates are less dangerous (in the economic sense) but may mean that a wind turbine with a larger generator should have been installed, with possible overall cost savings.

A knowledge of the aerodynamics of wind turbines is needed for two reasons. The first is to enable the aerodynamic performance of the machine to be assessed over the complete range of operating conditions. At its most basic, this requires determination of the torque exerted by the turbine blades on the rotor hub and shaft, and hence the available power. The second reason is to determine the loading and therefore the stresses on various parts of the wind turbine, so it can be designed to operate without failure.

This part of the book describes the characteristics of the wind, discusses the aerodynamics of wind turbines, and covers components and operational characteristics of common types of wind turbines.

1

The Wind Energy Resource

AIMS

This unit describes how the wind is caused by the heat of the sun falling on the earth's surface. It outlines the basic physical phenomena that are related to the characteristics of the wind. It gives formulas and definitions for simple calculations of wind statistics. It defines the energy content of the wind and explains how this can be converted into useful energy.

OBJECTIVES

At the end of this unit you will understand the main factors causing the wind and you will be able to do three things:

1. Carry out simple calculations of the energy flux in the wind, the wind speed frequency distribution and the vertical wind speed gradient.
2. Estimate the wind energy resource at a specific site.
3. Estimate the power and energy output from a typical wind turbine.

NOTATION AND UNITS

Symbol		Units
A	Area	m^2
b	Downstream velocity factor	
B	Atmospheric pressure	mm Hg
C	Scale parameter in a Weibull distribution	m/s
C_p	Coefficient of performance or power coefficient, a measure of rotor efficiency	
F	Force on rotor	N
k	Shape parameter in a Weibull distribution	

Symbol		Units
m	Mass flow rate	kg/s
p	Weibull probability density function	
P	Weibull cumulative probability distribution	
T	Air temperature	K
Tu	Turbulence intensity	
V	Wind speed	m/s
W	Power	J/s, W
z	Height above ground level	m
z_0	Roughness length	m
ρ	Air density	kg/m^3

Notes: The terms *velocity* and *speed* are used interchangeably throughout this book.

1.1 THE SOURCE OF THE WIND

The sun's energy falling on the earth produces the large-scale motion of the atmosphere, on which are superimposed, local variations caused by several factors. Due to the heating of the air at the equatorial regions, the air becomes lighter and starts to rise, and at the poles the cold air starts sinking. The rising air at the equator moves northward and southward. This movement ceases at about 30° N and 30° S, where the air begins to sink and a return flow of colder air takes place in the lowest layers of the atmosphere. The Coriolis acceleration due to the rotation of the earth causes the flow from the equator to the poles to be deflected towards the east, and the return flow towards the equator will be deflected towards the west producing the so-called trade winds.

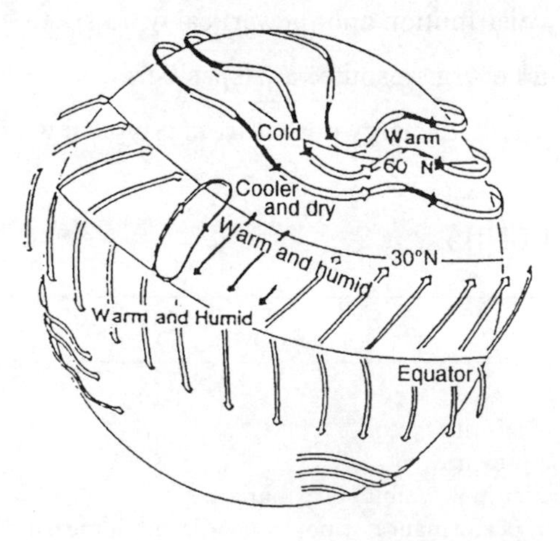

Figure 1.1. General circulation of winds over the surface of the earth. Reproduced by permission of the World Meterological Organisation

North of 30° N and south of 30° south the atmospheric motion is characterised by westerly winds. The general large-scale motion of the atmosphere is illustrated in Figure 1.1.

Differential heating of sea and land causes more minor changes in the flow of the air. The nature of the terrain, ranging from mountains and valleys to more local obstacles such as buildings and trees, also has an important effect on the wind.

1.2 THE CHARACTERISTICS OF THE WIND

The power in the wind is proportional to the cube of the wind speed or velocity, as will be shown below. It is therefore essential to have detailed knowledge of the wind and its characteristics if the performance of wind turbines is to be estimated accurately. As is well known, the highest wind velocities are generally found on hill tops, exposed coasts and out at sea. Various parameters need to be known of the wind, including the mean wind speed, directional data, variations about the mean in the short term (gusts), daily, seasonal and annual variations, and variations with height. These parameters are highly site specific and can only be determined with sufficient accuracy by measurements at a particular site over a sufficiently long period. They are used to assess the performance and economics of a wind plant. General meteorological statistics may overestimate wind speeds at a specific site.

The following sections touch on the complex processes which produce the wind in a particular location and they look in more detail at practical quantitative descriptions of wind characteristics.

1.3 THE ATMOSPHERIC BOUNDARY LAYER

The microscale flow is influenced by local features such as buildings and trees and the nature of the ground; e.g. whether a field is ploughed or has a standing crop. Areas of water such as lakes or the sea also have an important effect. The resulting changes in friction at the surface introduce fluctuations in the flow, and the wind velocity varies in time and space. The variation in wind velocity is illustrated in Figure 1.2, which shows a typical sample time history of wind speed measured by an anemometer mounted on the nacelle of a wind turbine. Some of the variation in wind speed is due to the turbulence created by the wind turbine rotor and the nacelle, but this does not hide the complex structure of the wind.

The instantaneous wind speed V can be described as a mean wind speed V_m plus a fluctuating wind component v:

$$V = V_m + v \tag{1.1}$$

The mean wind speed V_m is typically determined as a 10 minute average value.

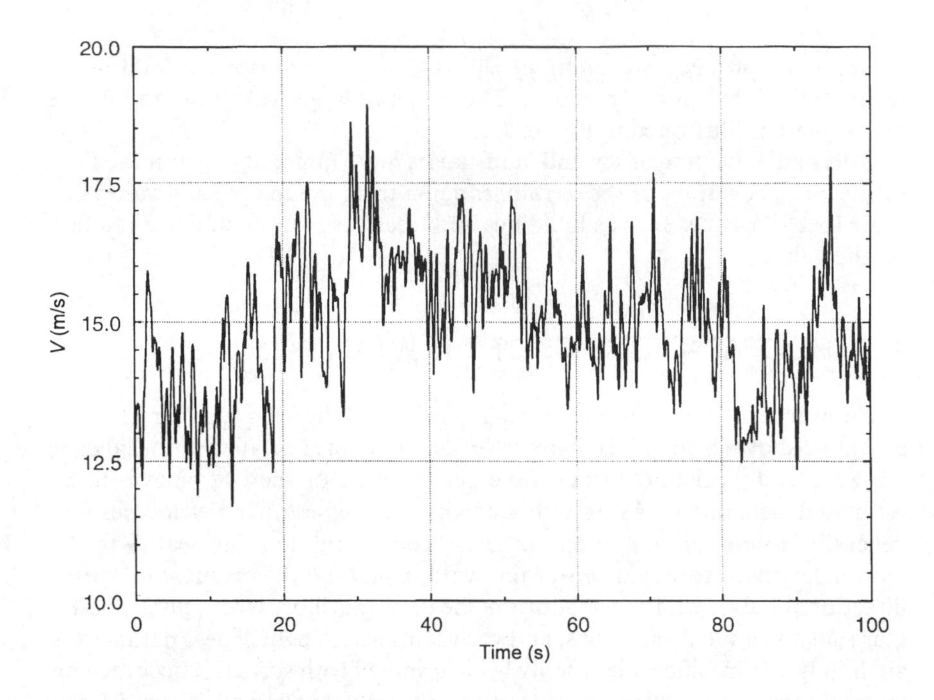

Figure 1.2. Wind speed measurements taken over 100s by the nacelle anemometer of a 33 m diameter wind turbine

The fluctuation of the flow is expressed in terms of the root mean square RMS value of the fluctuating velocity component ($\sqrt{\bar{v}^2}$) and is defined as the turbulence intensity given by

$$Tu = \frac{\sqrt{\bar{v}^2}}{V_{\mathrm{m}}} = \frac{1}{V_{\mathrm{m}}}\left[\frac{1}{T}\int_0^T v^2 dt\right]^{\frac{1}{2}} \tag{1.2}$$

For very rough terrain (e.g. where there are many trees and buildings) the turbulence intensity is in the range 0.15–0.2. For smooth terrain the intensity is typically 0.1.

1.4 VERTICAL WIND SPEED GRADIENT

The wind speed at the surface is zero due to the friction between the air and the surface of the ground. The wind speed increases with height most rapidly near the ground, increasing less rapidly with greater height. At a height about 2 km above the ground the change in the wind speed becomes zero. The vertical variation of the wind speed, the wind speed profile, can be expressed by different functions. Two of the more common functions which have been developed to describe the change in mean wind speed with height are based on experiments and are given below.

Power exponent function

$$V(z) = V_r \left(\frac{z}{z_r}\right)^{\alpha} \tag{1.3}$$

where z is the height above ground level, V_r is the wind speed at the reference height z_r above ground level, $V(z)$ is the wind speed at height z, and α is an exponent which depends on the roughness of the terrain.

A typical value of α might be 0.1. Thus if the wind speed at 10 m is 20 m/s, the wind speed at 40 m would be $20\left(\frac{40}{10}\right)^{0.1}$, which is 23 m/s.

Logarithmic function

$$\frac{V(z)}{V(10)} = \frac{\ln\left(\frac{z}{z_0}\right)}{\ln\left(\frac{10}{z_0}\right)} \tag{1.4}$$

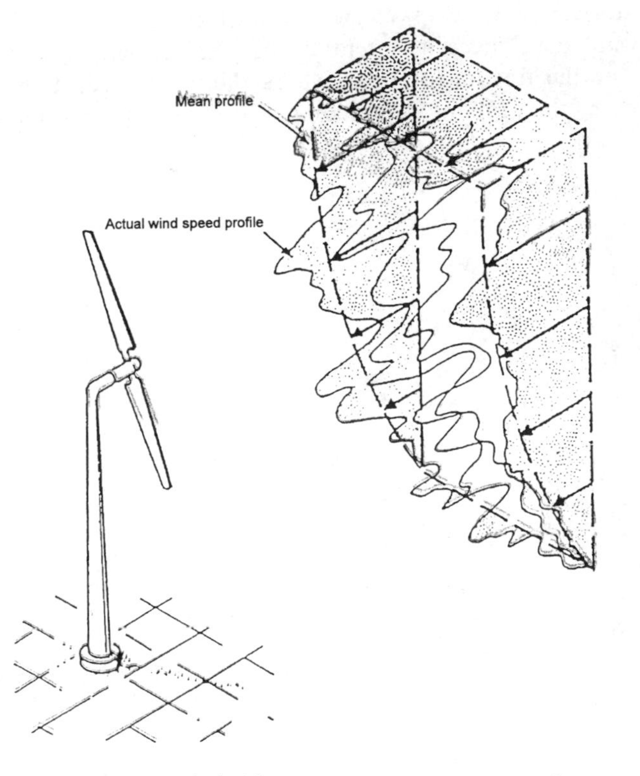

Figure 1.3. Representation of wind flow in the boundary layer near the ground

where $V(10)$ is the wind speed at 10 m above ground level and z_0 is the roughness length.

The parameters α and z_0 for different types of terrain are shown in Table 1.1.

Table 1.1. Wind speed: parameters for calculating a vertical profile

Type of terrain	Roughness class	Roughness length, z_0 (m)	Exponent, α
Water areas	0	0.001	0.01
Open country, few surface features	1	0.12	0.12
Farmland with buildings and hedges	2	0.05	0.16
Farmland with many trees, forests, villages	3	0.3	0.28

Both functions can be used for calculation of the mean wind velocity at a certain height, if the mean wind velocity is known at the reference height.

An illustration of the flow in the boundary layer is given in Figure 1.3.

1.5 WIND STATISTICS

A typical histogram of the wind velocity is shown in Figure 1.4. Derived from long-term wind data covering several years, the histogram indicates the probability, or the fraction of time, where the wind speed is within the interval given by the width of the columns. The data was measured in 1 m/s

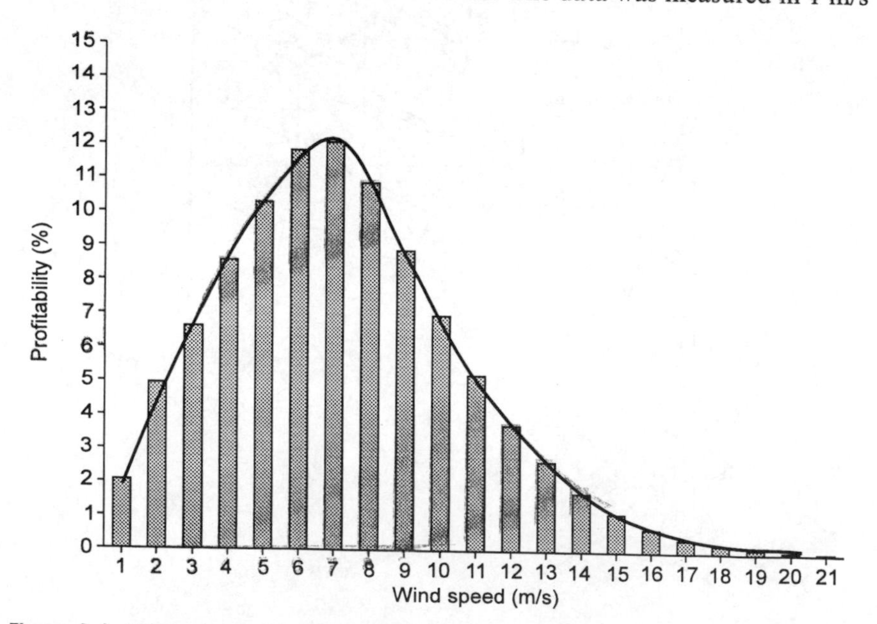

Figure 1.4. Histogram and Weibull function for the probability of a given wind speed (data measured in 1 m/s bins)

bins i.e. 4.5–5.5, 5.5–6.5 m/s, etc. The sum of the height of the columns is 1 or 100%. When the width of the columns becomes smaller, the histogram becomes a continuous function called the probability density function. A simple and useful interpretation of the wind speed probability density function is that it shows the probability of a wind speed being in a 1 m/s interval centred on a value of V. Thus, referring to Figure 1.4, the probability of the wind speed being between 4.5 and 5.5 m/s is 0.104 or $(0.104 \times 8760) = 910$ hours per year.

The histogram takes into account the seasonal variation and the year-by-year variation for the years covered by the statistics.

The probability density function can be fitted to a *Weibull function* given by

$$p(V) = \frac{k}{V}\left(\frac{V}{C}\right)^{k-1} \exp\left\{-\left(\frac{V}{C}\right)^{k}\right\} \tag{1.5}$$

where $p(V)$ is the frequency of occurrence of wind speed V, C is the scale parameter or characteristic wind speed, and k is the shape parameter.

The cumulative Weibull distribution $P(V)$ gives the probability of the wind speed exceeding the value V; it is expressed as

$$P(V) = \exp\left\{-\left(\frac{V}{C}\right)^{k}\right\} \tag{1.6}$$

A typical value for k would be 2; when $k = 2$ the distribution is called a cumulative Rayleigh distribution.

Figures 1.5 and 1.6 are cumulative distributions which show the effect of two different scale parameters, 10 m/s and 5 m/s respectively.

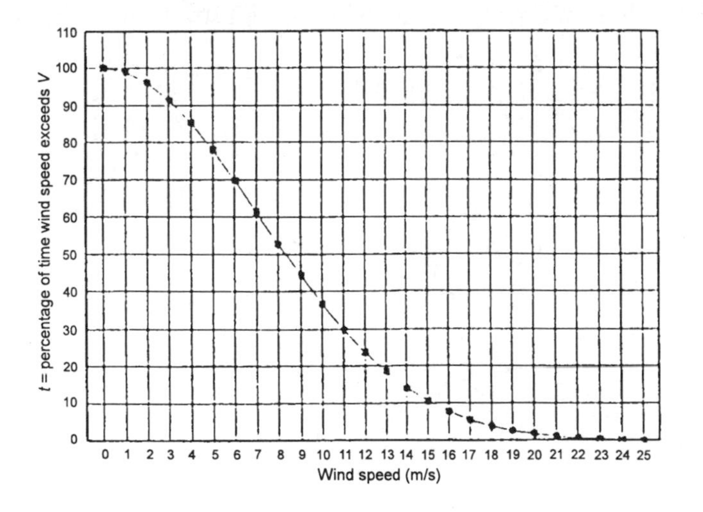

Figure 1.5. Weibull cumulative distribution, shape parameter $C = 10$ m/s

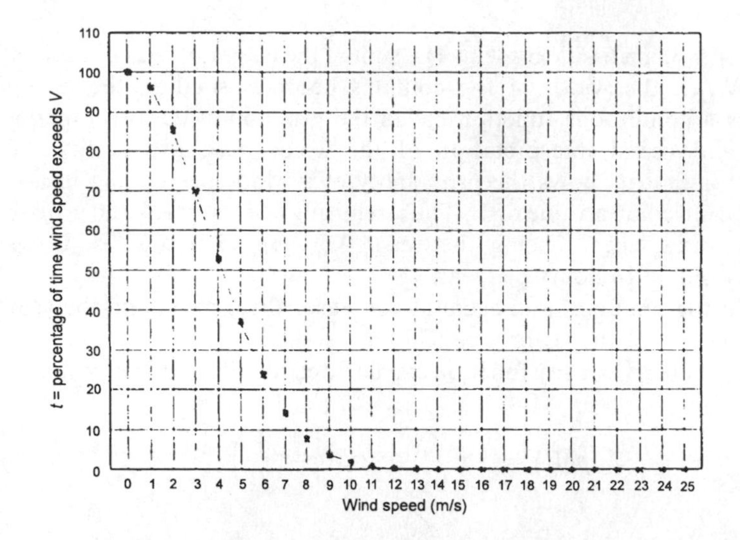

Figure 1.6. Weibull cumulative distribution, shape parameter C = 5 m/s

The probability of a wind speed between V_1 and V_2 is given by

$$P(V_1 < V < V_2) = \exp\left\{-\left(\frac{V_1}{C}\right)^k\right\} - \exp\left\{-\left(\frac{V_2}{C}\right)^k\right\} \tag{1.7}$$

The parameters C and k for the Weibull frequency distribution can be found by plotting $\ln V$ against ln(-ln($P(V)$)), where ln is the logarithm to base e, and fitting a straight line to the points. The slope of the line is equal to k and C is equal to exp($\ln V$), or V, where ln(-ln($P(V)$)) is zero. This technique is based on taking logarithms of Equation (1.6) twice. It is illustrated in Figure 1.7.

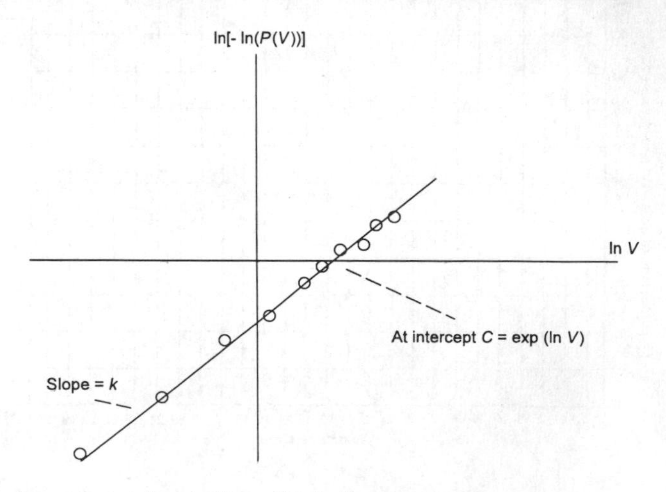

Figure 1.7. Graphical determination of Weibull parameters

1.6 ENERGY IN THE WIND

We need to begin with two fundamental concepts *power*, or energy per unit time, and *energy*, which is available over a given period of time. The kinetic energy in a flow of air through a unit area perpendicular to the wind direction is $\frac{1}{2}V^2$ per unit mass or $\frac{1}{2}\rho V^2$ per unit volume.

For an airstream flowing through an area A the mass flow rate is ρAV, therefore

$$\text{Power, } W = (\rho AV)1/2V^2 = 1/2\rho AV^3 \quad (1.8)$$

where ρ is the air density (kg/m^3), V is the wind speed (m/s) and W is the power (watts or joules/per second). The power is also known as the energy flux or power density of the air.

The air density ρ is a function of the air pressure and the air temperature:

$$\rho = \rho_0\left(\frac{288B}{760T}\right) \quad (1.9)$$

where ρ_0 is the density of dry air at standard temperature and pressure (1.226 kg/m^3 at 288 K, 760 mm Hg), T is the air temperature (K) and B is the barometric pressure in millimetres of mercury.

Both the pressure and the temperature are functions of the height above sea level. Taking a typical density of air at sea level as 1.2 kg/m^3, the power becomes

$$W = 0.6V^3 \text{per unit area} \quad (1.10)$$

At a wind speed V the *energy* measured in watt-seconds passing through area A during time t is given by

$$\text{Energy, } Wt = 1/2\rho AV^3t \quad (1.11)$$

This is the total energy available for doing work on the wind turbine. Note that, practically speaking, there is no change in the temperature of the air flowing through a wind turbine. This is also clearly the case in water turbines. In both cases the energy is extracted using the change in fluid velocity, not through a change in temperature. This would not be the case in a gas or steam turbine.

The Betz limit

A mathematical derivation of the Betz limit may be considered optional, but be aware of the Betz limit and the resulting maximum value of the coefficient of performance C_p of 59%

The power W is the total *energy per unit time* in the airstream, only a proportion of which can be converted to useful energy by a wind turbine. The power available for a wind turbine is equal to the change in kinetic energy of the air as it passes through the rotor. Consider the flow through a simple stream tube enclosing the rotor (Figure 1.8).

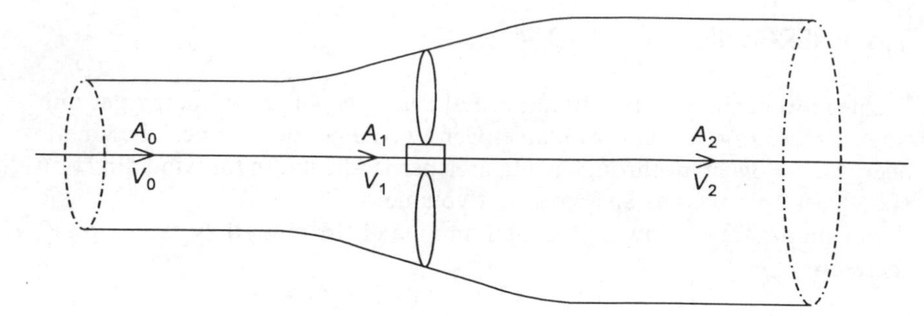

Figure 1.8. Ideal air flow through a wind turbine

The airflow can be regarded as incompressible; the air velocities are normally low, so density changes can be treated as negligible. The streamlines must diverge as they pass through the rotor disc as the velocity reduces, and in the simple momentum theory considered here the pressures immediately upstream and downstream of the rotor are constant over the rotor disc. This in effect assumes that the two or three blades typical of a modern wind turbine have been replaced by an infinite number of thin blades, giving the same force on the approaching airflow as the average for the real rotor system.

The mass flow rate is constant far upstream (0), at the rotor (1), and far downstream (2).

$$\text{Mass flow rate, } m = \rho A_0 V_0 = \rho A_1 V_1 = \rho A_2 V_2 \tag{1.12}$$

The force F on the rotor disc is given by the rate of change of momentum

$$F = m(V_0 - V_2) \tag{1.13}$$

The power W extracted by the rotor is given by the rate of change of kinetic energy

$$W = m(1/2V_0^2 - 1/2V_2^2) \tag{1.14}$$

At the rotor, the force does work at speed V_1 so that

$$W = FV_1 \tag{1.15}$$

Using equations (1.13 to 1.15) and with a little manipulation we can show that

$$V_1 = 1/2(V_0 + V_2) \tag{1.16}$$

We now introduce the downstream velocity factor, b, to describe the ratio of upstream to downstream wind speeds,

$$b = \frac{V_2}{V_0} \tag{1.17}$$

and using equations (1.15), (1.14) and (1.12) it can be seen that

$$\frac{F}{A_1} = 1/2\rho V_0^2(1 - b^2) \tag{1.18}$$

then using equations (1.15) and (1.16)

$$\frac{W}{A_1} = 1/2\rho V_0^3 \times 1/2(1 - b^2)(1 + b) \quad (1.19)$$

The coefficient of performance C_p is defined as the fraction of energy extracted by the wind turbine of the total energy that would have flowed through the area swept by the rotor if the turbine had not been there.

$$C_p = \frac{W}{W_1} \quad (1.20)$$

and since

$$W_1 = 1/2\rho A_1 V_0^3 \quad (1.21)$$

we have

$$C_p = 1/2(1 - b^2)(1 + b) \quad (1.22)$$

Differentiation of C_p with respect to b shows C_p to be a maximum when the speed ratio b is 1/3, giving

$$C_p = \frac{16}{27} \text{ or about } 59\%$$

This is the Betz limit, first formulated in 1919, and applies to all types of wind turbines. Intuitively there must be a wind speed change at which the conversion efficiency is a maximum. If there were no change in wind speed, no energy would be extracted and the power of the wind turbine would be zero. If the air were brought completely to rest, all its energy would dissipated. However, a rotating wind turbine will not completely prevent the flow of air, so it can only extract a proportion of the kinetic energy in the wind. The wind speed onto the rotor at which energy extraction is a maximum therefore lies somewhere between the upstream speed and zero wind speed.

Modern designs of wind turbines for electricity generation operate at C_P values of about 0.4. The major losses in efficiency for a real wind turbine arise from the viscous drag on the blades, the swirl imparted to the airflow by the rotor, and the power losses in the transmission and electrical system.

1.7 CONVERSION OF WIND ENERGY

How can the variations in the wind velocity over a period of time be converted into an annual energy output in kilowatt-hours (kWh) for the wind turbine?

It should be clear by now that we need to start from a time series of wind velocity, from which we can integrate the power in a series of small time intervals over the specified period. This can be done from the turbine power characteristic with the cumulative distribution of wind speeds. Account needs to be taken of the variation of wind speed with height, so a value representative of the speed at the hub has to be chosen.

Depending on local wind speeds, a turbine will produce an annual average power that is some proportion of its maximum rated power, typically up to 30%. Assuming the wind turbine will be available to operate for 95% of the time, the overall annual *load factor* or *capacity factor* would then be 0.3×0.95 or 28.5%.

A 100 kW wind turbine at full power throughout the year would produce 8.76×10^5 kWh per year, its theoretical maximum. The 100 kW is only a fraction of the kinetic energy of the wind significantly lower than the Betz criterion and perhaps as low as 0.4. At an annual capacity factor of 28.5% the actual output would be 2.5×10^5 kWh per year, which at 10 cents per unit would be worth $25 000. The ideal power characteristic for a wind turbine is shown in Figure 1.9.

Below 5 m/s, the *cut-in wind speed*, there is not enough energy in the wind to overcome the mechanical losses within the turbine. Between 5 m/s and 13 m/s the output power rises rapidly with wind speed and is given by

$$W = C_p \times 1/2\rho A_1 V_0^3 \tag{1.23}$$

Equation (1.23) does not actually produce a true cubic relationship of output power with wind speed; this is because the coefficient of performance, C_p, which is a measure of the efficiency of the turbine, varies with wind speed.

In this example the designer of the wind turbine has decided to limit the power converted by the rotor to 100 kW. This design choice was made in order to limit the strength and therefore the weight and cost of the components of the turbine. Over the year some energy will be lost because of this decision. But, consider the typical wind speed distribution of Figure 1.4. Notice that the number of hours per year is quite small when the wind speed exceeds 13 m/s, so the wind power available to the turbine rotor exceeds

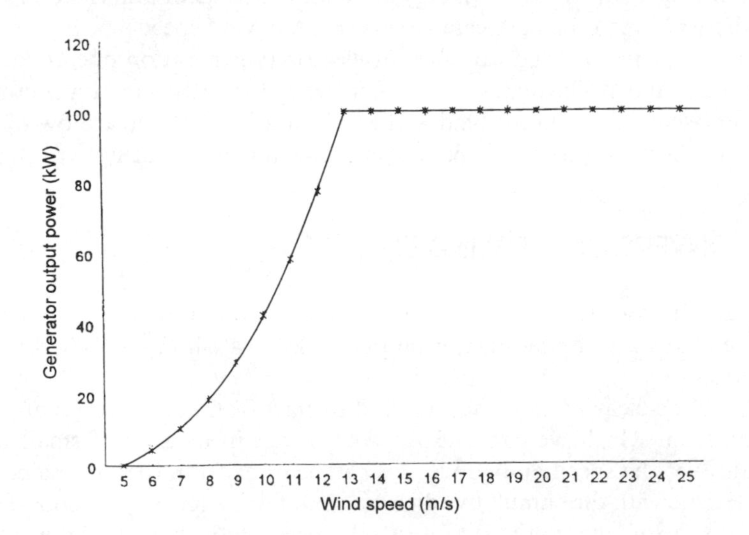

Figure 1.9. Theoretical characteristic for a 100 kW wind turbine

100 kW. The annual energy lost will therefore be small, although at sites where the wind speed is high a different maximum power, or rating, might have been selected.

There are two main techniques for limiting or *regulating* the power of a wind turbine. *Pitch regulation* rotates the wind turbine blades mechanically in order to reduce their aerodynamic efficiency, thereby lowering C_p. *Stall regulation* allows the blades to go into a condition of aerodynamic stall when the wind speed is high; the blades are not mechanically moved. Stall regulation also reduces aerodynamic efficiency, limiting the power which must be transmitted by the drivetrain to the generator.

The wind speed, at hub height, when the turbine produces its maximum (or rated) power is known as its *rated wind speed* and is 13 m/s in this example.

In practice the power curve of a wind turbine is measured using 10 minute mean values of wind speed and output power. The wind speed can only be measured at some distance from the turbine and cannot account for variations in wind speed across the rotor disc. Thus a large number of 10 minute measurements are taken and averaged using a binning technique. This produces a power curve with a rather more rounded shape than the curve in Figure 1.9.

1.8 THE WORLD WIND RESOURCE

To estimate the global potential for wind energy it is necessary to know the mean wind speed over the earth's surface. This has been measured and the results are published in the form of wind atlases for countries and regions, even for the earth as a whole. Although these atlases often depend on interpolation of wind data obtained from dispersed measuring stations, they can give a useful estimate of the available wind resource for initial planning purposes.

The World Energy Council (1994) gives estimates of the global wind resource. About 27% of the earth's land surface experiences annual mean wind speeds higher than 5.1 m/s at 10 m above the surface. Only 4% of this area might be available for electricity-generating wind farms because of unsuitable terrain, urban areas, crop cultivation and other existing land use. Assuming a generating capacity of 8 MW/km^2 and a capacity factor of 23%, it is estimated that the global potential for wind turbine power production is 20 000 TWh per year. For comparison in 1987 the total world electricity consumption was about 10 500 TWh (the prefix T stands for tera or 10^{12}).

Note that this estimate is for large scale grid-connected wind turbines and it depends on a variety of assumptions. It does not include the potential for offshore wind energy developments or small-scale wind turbines used for water pumping or battery charging, feasible applications in wind speeds as low as 3 m/s. About 50% of the earth's land surface is exposed to mean annual wind speeds of between 4.4 m/s and 5.1 m/s quite suitable for small wind turbines.

In 1991 the International Energy Agency estimated that the total world installed capacity of grid-connected wind turbines was 2200 MW, amounting to 21 000 machines generating about 3.8 TWh, assuming a capacity factor of 20%. By 1996 the total installed capacity of grid-connected wind turbines had increased to over 5000 MW. Over one million windpumps are thought to be in operation with an energy production of over 150 MWh per year. Notice that only a small proportion of the feasible wind energy potential has been utilised so far. Some of the constraints on wind power development are discussed in Chapters 4 and 5. The World Energy Council (1994) have estimated that by the year 2020, and if current policies continue, wind power could provide 375 TWh per year to the world electricity supply, about 1.5% of the forecast world electricity consumption. If incentives of various kinds were provided, perhaps through an ecological imperative, this estimate might become higher by a factor of 3. It is therefore likely that the use of the wind as a source of power will grow substantially in the coming decades.

SUMMARY OF THE CHAPTER

A description was given of how the wind is produced by the heat of the sun. The basic physical phenomena relating to the characteristics of the wind were reviewed, alone with some formulas and definitions relevant to wind statistics. The importance of site-specific wind statistics was discussed. The energy content of the wind was defined and it was described how this can be converted into useful energy using a typical modern wind turbine. The theoretical limit to energy conversion was derived. Some current estimates were given for the world wind energy resource and its future development.

BIBLIOGRAPHY AND REFERENCES

International Energy Agency, *Wind Energy Annual Report 1991*, Paris, 1992.

Solar Energy Research Institute, *The Potential for Wind Energy in Developing* Countries, Report SERI/STR-217-3219, SERI, Golden, CO, 1987.

World Energy Council, *New Renewable Energy Resources*, Kogan Page, London, 1994.

World Meteorological Organisation, *Meteorological Aspects of the Utilization of Wind as an Energy Source*, Technical Note 175, WMO, Geneva, 1981.

SELF-ASSESSMENT QUESTIONS

PART A. True or False?

1. The trade winds are due to the Coriolis acceleration caused by the rotation of the earth.

2. The power in the wind is proportional to the square of the wind speed.

3. Near the ground the wind speed does not vary with height.

4. The power of the wind is the energy measured over a period of time.

5. The energy extracted by a wind turbine arises primarily from a change in air velocity.

6. The coefficient of performance measures the fraction of the energy extracted by a turbine from the wind.

7. A wind turbine could be designed to extract over two-thirds of the available wind energy.

8. If the wind velocity were zero behind the rotor of a wind turbine, its coefficient of performance would be 100%.

9. The power developed by a real wind turbine varies with time.

10. If a wind turbine could operate throughout the year, its annual average power would be the same as its rated power.

PART B

1. What causes the major motions of the air in the earth's atmosphere?

2. What are some of the minor causes of the variations in the wind?

3. What is the typical shape of the time-averaged profile of the wind speed from ground level to several kilometres high?

4. Name the two functions that are commonly used to calculate the variation of wind speed with height from the ground.

5. What does the Weibull function describe?

6. What is the practical use of the Betz limit for power production of a wind turbine?

7. A horizontal-axis wind turbine has a rotor diameter of 20 m, a rotor hub height of 20 m and reaches its full-load electrical output at the generator terminals of 100 kW at a wind speed of 13 m/s. The air density is 1.2 kg/m^3. Assume an empirical power-law relationship between wind speed and height, where the exponent in the power law is 0.13. If the turbine has just reached its rated electrical output, calculate the following quantities to three significant figures:
(a) the wind speed at 10 m
(b) the wind speed at 20 m
(c) the wind speed at 30 m

8. For the wind turbine in Question 7, what is the power of the wind passing through the area swept by the rotor.

9. For the wind turbine in Question 7, what is the coefficient of performance C_p of the wind turbine at a wind speed of 13 m/s; give your answer to three significant figures.

10. In Question 9 what fraction of the theoretical maximum C_p is the coefficient of performance of this turbine at a wind speed of 13 m/s?

Answers

Part A

1. True; 2. False; 3. False; 4. False; 5. True; 6. True; 7. False; 8. False; 9. True; 10. False.

Part B

1. The sun's energy causing the warm air to rise at the equatorial regions and the cold air to fall at the polar regions. The earth's rotation then causes the moving layers of air to be deflected eastwards or westwards according to their position.

2. Differential heating between sea and land, mountains and valleys, roughness of terrain, artificial structures.

3. See Figure 1.3.

4. The power exponent function and the logarithmic function.

5. The Weibull function is an empirical fit to the probability density function of the time history of the wind at a particular location. It gives the probability or the fraction of time that the wind speed has a particular magnitude. Note that the cumulative Weibull distribution function gives the probability of the wind speed exceeding a particular value.

6. To indicate the efficiency a real wind turbine, its coefficient of performance should be compared with the Betz limit of 59.3%, not with 100%.

7. Full power is reached at 13 m/s, which is the rated wind speed for the turbine at hub height; then (a) 11.9 m/s (b) 13.0 m/s (c) 13.7 m/s.

8. 414 kW

9. The actual output of the turbine is 100 kW, so $C_p = 100/414 = 0.242$.

10. The fraction is $0.242/0.593 = 0.408$.

2

Aerodynamics

AIMS

This unit discusses some of the aerodynamic ideas underlying the operation of wind turbines. It considers the basic rotor types, including the differences between lift and drag type rotors, vertical and horizontal axis wind turbines, and fixed and variable speed operation. After outlining the basic aerodynamic models relating to wind turbine design, it moves on to introduce blade element theory.

OBJECTIVES

When you complete this unit you will be able to do three things:

1. Appreciate the differences between a variety of wind turbines.
2. Understand the underlying principles of wind turbine aerodynamics.
3. Know the procedure for the initial aerodynamic analysis of a horizontal-axis rotor.

NOTATION AND UNITS

Symbol		Units
a	axial velocity interference factor	
a'	Radial velocity interference factor	
A	Area	m^2
B	Number of blades	
c	Aerofoil chord length	m
C_D	Drag coefficient	
C_L	Lift coefficient	
C_M	Moment coefficient	
C_p	Coefficient of performance or power coefficient, a measure of rotor efficiency	

Symbol		Units
D	Drag force	N
D	Rotor diameter	m
F	Force on blade element	N
M	Pitching moment	N m
Q	Torque	N m
R	Rotor radius	m
T	Thrust	N
V	Wind speed	m/s
W	Power	J/s, W
X	Tip speed ratio	
α	Angle of incidence, angle of attack	
ρ	Air density	kg/m^3
σ	Blade solidity	
ω	Rotational speed	rad/s

2.1 ROTOR TYPES

In principle there are two different types of wind energy conversion devices: those which depend mainly on aerodynamic *lift* and those which use mainly aerodynamic *drag*.

Low speed devices are mainly driven by the drag forces acting on the rotor. They generally move slower than the wind, and their motion reduces rather than enhances the power extraction. The torque at the rotor shaft is relatively high.

High speed turbines rely on lift forces to move the blades, and the linear speed of the blades is usually several times faster than the wind speed. The torque is low compared to the drag type.

The main characteristics for different types of wind turbines are summarised in Figure 2.1. This illustrates how some traditional windmills and some pumping devices are of the low speed drag type, whereas modern turbines for electricity production are of the high speed lift type. Given the same swept area, the power extracted by a wind turbine relying on lift forces is generally many times greater than the power from a turbine relying on drag. For the generation of electricity it is usually desirable that the driving shaft of the generator operates at a considerable speed (1000 or 1500 revolutions per minute). This, together with the much higher aerodynamic efficiency of lift type devices, means that devices relying on aerodynamic drag are not normally used for electricity generation.

Wind turbines can also be classified into horizontal-axis and vertical-axis machines, although the Betz limit for energy conversion applies to any type of wind turbine. *Horizontal-axis*, or propeller-type, turbines are more common and highly developed than the vertical-axis designs and their analysis is continued in the next sections.

Rotor Type	*Typical load*	*(rpm)*	C_p	*Torque*
Propeller (lift)  Double-bladed  Three-bladed	Electrical generator	high	0.42	low
Darrieus (lift) 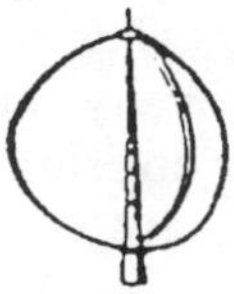	Electrical generator	high	0.40	low
Cyclogiro (lift) 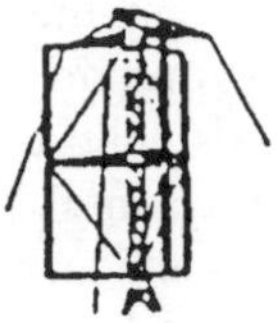	Electrical generator or pump	moderate	0.45	moderate
Chalk Multiblade (lift) 	Electrical generator or pump	moderate	0.35	moderate
Sailwing (lift) 	Electrical generator or pump	moderate	0.35	moderate
Fan-type (drag) 	Electrical generator or pump	low	0.30	high
Savonius (drag) 	pump	low	0.15	high
Dutch-type (drag) 	pump of millstone	low	0.17	high

Figure 2.1. Operating characteristics for different rotor types. (Reproduced from Technical note 175 by permission of the World Metereological Organisation)

Vertical-axis wind turbines range from the drag type, such as the Savonius type and cup anemometer for measuring wind speed, to high speed turbines where the blades are vertical and straight with a symmetrical aerofoil profile or are curved in the classic troposkien shape. The latter are usually known as the Darrieus type and the shape is such that the centrifugal loads are balanced by pure tension forces in the blades, thus avoiding bending moments. Modern vertical-axis machines have several advantages: they operate independently of the wind direction, so a yawing mechanism is not required; heavy gearboxes and generating machinery may be situated at ground level; and as they rotate, the blades do not suffer fatigue stresses from gravitationaly induced forces. They also have come disadvantages: vertical-axis machines are not self-starting; the torque fluctuates with each revolution as the blades move into and away from the wind; and speed regulation in high winds can be difficult. One or two organisations are continuing to investigate and develop vertical-axis wind turbines but so far they have not proved as cost-effective as their horizontal-axis counterparts. The commercial significance of vertical-axis wind turbines is therefore somewhat limited at present. A fuller description and aerodynamic analysis of vertical-axis wind turbines is given by Freris (1990).

A final distinction is whether the rotor is allowed to run at *variable speed* or constrained to operate at a *constant speed.* For water pumping and small battery charging operation it is desirable to allow the rotor speed to vary. However, for the large-scale generation of electricity it is common to operate wind turbines at constant speed. This allows the use of simple generators whose speed is fixed by the frequency of the electrical network. Variable-speed wind turbines are sometimes used for electricity generation but a power electronic frequency converter is then required to connect the variable-frequency output of the wind turbine to the fixed frequency of the electrical system.

There are several advantages in operating wind turbines at variable speed; the most obvious is an increase in aerodynamic efficiency. This can be seen clearly if the performance coefficient, C_p, of a rotor is plotted against the tip speed ratio, X.

The tip speed ratio, X, of a rotor is defined as

$$X = \frac{V_{tip}}{V_{wind}} = \frac{\omega R}{V_0} \tag{2.1}$$

where R is the rotor radius, ω the rotational speed in rad/s and V_0 the undisturbed wind speed.

From Chapter 1 we recall that the coefficient of performance, C_p, is defined as

$$C_p = \frac{W}{1/2 \rho A V_0^3} \tag{2.2}$$

It is conventional to plot the variation of the performance coefficient (or power coefficient), C_p, against the tip speed ratio, X, rather than against the wind speed, as this produces a dimensionless graph.

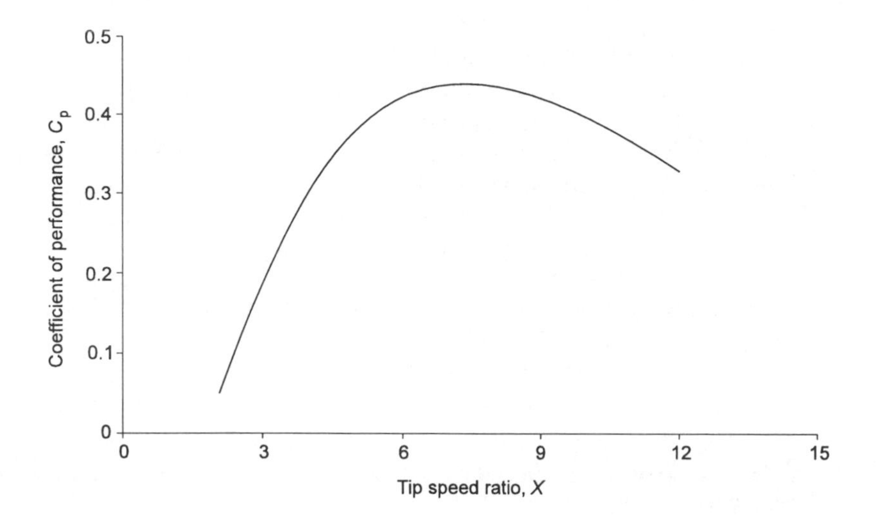

Figure 2.2. Curve showing variation of coefficient of performance with tip speed ratio

A typical C_p–X curve for a large two-bladed wind turbine rotor is shown in Figure 2.2.

Notice that the maximum value of C_p is only achieved at a particular tip speed ratio (approximately 7 in this example). For a fixed-speed wind turbine, where ω is constant, this corresponds to a particular wind speed. At all other wind speeds the efficiency of the rotor is reduced. Thus, it would seem desirable to operate at a constant tip speed ratio which, with varying wind speeds, implies that the rotor rotational speed must also vary. In practice the increase in energy capture possible with variable-speed operation is quite small and often exceeded by the electrical losses incurred in the variable-frequency conversion equipment. However, a very significant benefit of variable-speed operation of large wind turbines is that, by allowing the rotor to act as a large flywheel, it reduces the mechanical loads on the wind turbine drivetrain.

2.2 FORCES DEVELOPED BY WIND TURBINE BLADES

As a first approach the aerodynamic forces acting on a wind turbine rotor can be explained by classic aerofoil theory (Glauert, 1959).

When the aerofoil moves in a flow, a pressure distribution is established around the aerofoil. This distribution is a function of the angle of incidence, which is the angle between the chord and the direction of the flow. On the upper side of the aerofoil there is a negative pressure (suction); on the lower side there is a positive pressure. Figure 2.3 shows how the loads on the aerofoil due to the pressure distribution can be represented as two forces and one torque.

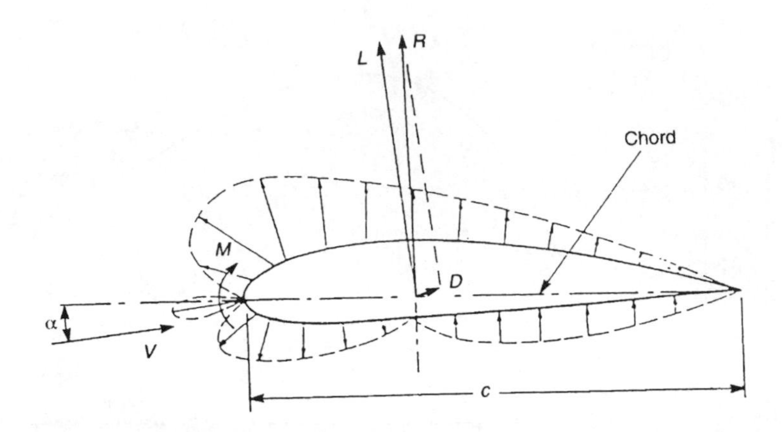

Figure 2.3. Pressure distribution around an aerofoil, resulting lift and drag forces L and D, and pitching moment M

The lift L is perpendicular to the direction of the airflow V; the drag D is parallel to the flow direction and perpendicular to the lift. The torque M is called the pitching moment; it is given relative to a specific point, usually at 0.25 of the chord.

The lift, drag and moment are usually expressed as dimensionless aerodynamic coefficients of the aerofoil, defined as follows:

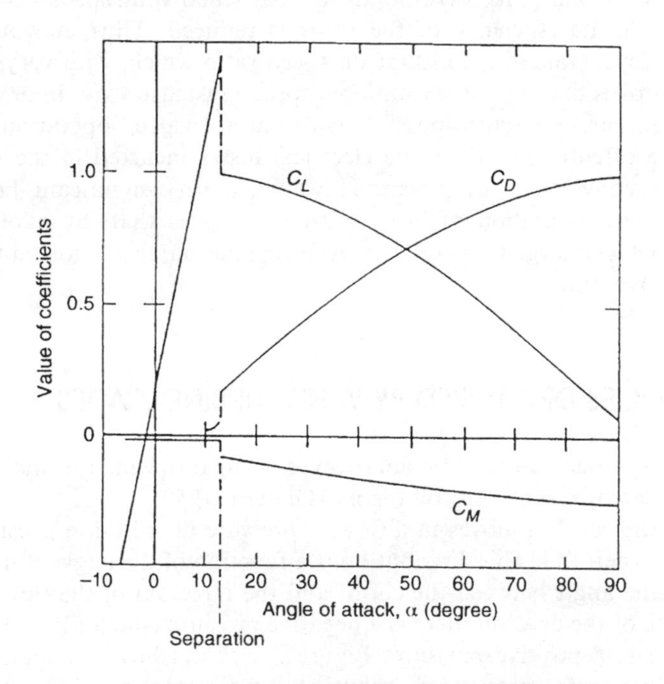

Figure 2.4. Typical variation of aerofoil coefficients with angle of incidence

$$C_L = \frac{L}{1/2\rho A V^2} \quad C_D = \frac{D}{1/2\rho A V^2} \quad C_M = \frac{M}{1/2\rho c A V^2} \tag{2.3}$$

where ρ and V are the air density and wind speed, and A and c are the plan area and chord length of the aerofoil.

The coefficients are functions of the angle of incidence and typical variation of C_L, C_D and C_M are shown in Figure 2.4. The sudden change in the coefficients at approximately 13° is due to flow separation from the suction side of the aerofoil; this is called *stall*. The coefficients also depend on the Reynolds number, based on the chord and upstream relative velocity, which for large wind turbines is of the order of 2×10^6. Values of coefficients for a wide range of aerofoils are given in Abbott and Von Doenhoff (1959).

In a wind turbine rotor the lift and drag forces on the blades are transformed into a rotational torque and an axial thrust force. The torque produces useful work whereas the thrust will try to overturn the turbine and must be resisted by the tower and foundations. Figure 2.5 shows how a single element of the blade may be considered and, by viewing the element from beyond the end of the blade, the wind speed and forces on the element can be shown as in Figure 2.6.

The lift and drag forces depend on the characteristics of the aerofoil, the magnitude of the relative wind speed and the angle it makes with the chord

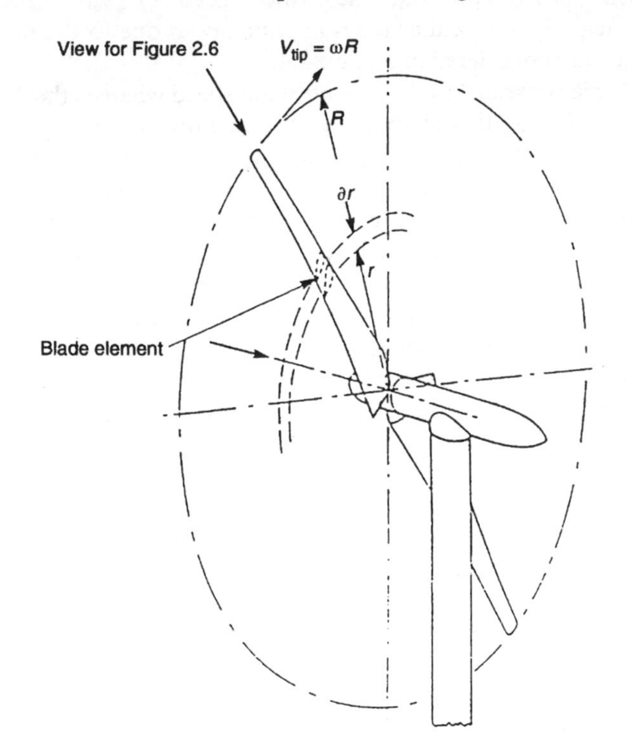

Figure 2.5. Radial blade element on a rotor blade

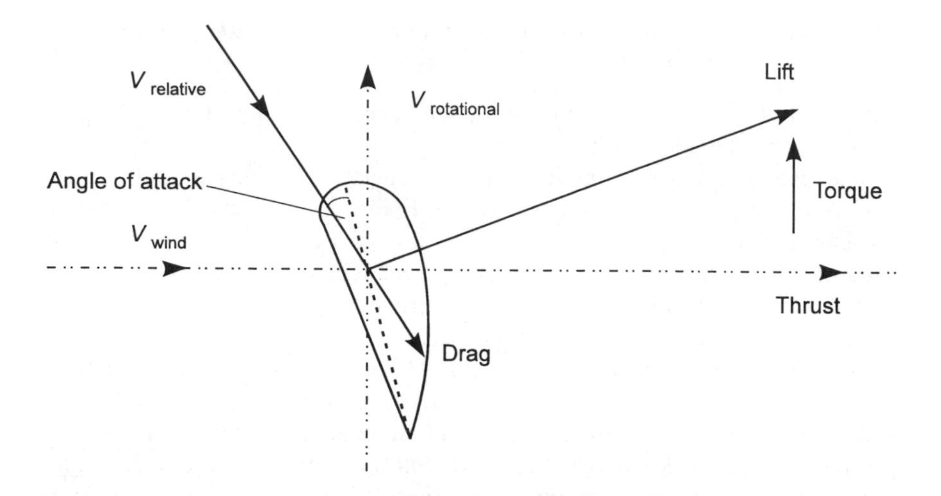

Figure 2.6. Simple representation of the forces on a blade element

line of the aerofoil (the angle of attack). The relative wind speed (V_{relative}) is found from the triangle of velocities created by the rotational speed of the blade element ($V_{\text{rotational}}$) and the wind speed (V_{wind}). This simple explanation neglects the local changes in wind speed due to the presence of the rotor; they are considered in section 2.4.

The drag force is parallel to the relative wind speed whereas the lift force is perpendicular. If the lift and drag forces are now resolved parallel and perpendicular to the direction of rotation of the blade element, the lift force creates useful torque but the drag force opposes it. Both lift and drag forces contribute to the axial thrust force on the rotor. The total useful torque developed by the rotor is simply the sum of the torques developed by each individual blade element; the same goes for the thrust force.

A high ratio of C_L to C_D is desirable for a high efficiency rotor and most of the power from a rotor is derived from the outer part of the blades, where the rotational speed of the element is greatest. This leads to a high torque operating over a long lever arm to the hub. Finally, if it is desired to keep the angle of attack constant over the length of the blade, it is necessary to build a twist into the blade shape. Blade twist is included in some designs, although this increases the complexity of their manufacture.

The mechanisms for controlling or regulating power at high wind speeds can now be understood. In the power curve of Figure 1.7 the output power of the turbine was limited to 100 kW for wind speeds above 13 m/s. In a fixed-speed turbine, limiting power is equivalent to limiting torque. This can be achieved by reducing the lift on the blade or by increasing drag.

Consider Figure 2.4 and suppose the blade has an angle of attack of 10°. This corresponds to a value of C_L of approximately 1 and a very low value of C_D.

In a *pitch-regulated* rotor, the blade (or an outer part of it) is physically rotated on bearings to reduce the angle of attack, α. Reducing the angle of attack reduces C_L; it reduces the lift force and there is a consequent reduction in torque and power.

Stall regulation is a most elegant solution to the problem of regulating power as it requires no moving parts. The blade is rigidly fixed to the hub. Again, consider Figure 2.4 and suppose the angle of attack is 10°. The wind turbine is of the fixed-speed type, so the rotational speed of the element ($V_{\text{rotational}}$) is constant. As the wind speed (V_{wind}) increases, the triangle of velocities indicates that the relative wind speed (V_{relative}) moves nearer to the wind speed (V_{wind}) and the angle of attack will increase. Once the angle of attack exceeds 13° the blade element enters the stall region. Beyond 13° C_L drops and C_D increases. Both of these changes lead to a reduction in torque and hence power. One difficulty of stall regulation is that, if the connection with the grid is lost, the wind turbine will overspeed and $V_{\text{rotational}}$ will increase. In this event the angle of attack is reduced, the blade comes out of stall and additional torque is developed by the rotor, thus accelerating the overspeed. So even more care is needed in the design of braking systems for stall-regulated wind turbines than for pitch-regulated designs.

The simple explanation of the operation of horizontal-axis wind turbines given in this section is intended only as an introduction to the subject. Real wind turbines have rather more complex aerodynamics and certain aspects, e.g. operation in the stall region, remain poorly understood.

2.3 AERODYNAMIC MODELS

The simplest model for wind turbine aerodynamics is the Rankine–Froude *actuator disc model*, used in section 1.6 to derive the Betz limit for energy conversion from the wind. The rotor is approximated by an actuator disc, which has an infinite number of thin dragless blades rotating with a tip speed higher than the wind speed. The model assumes a sudden drop in pressure at the rotor disc; this represents the force required to slow the wind and extract energy. The flow velocity is uniform in the radial direction and varies smoothly from far upstream to far downstream. The rotation of the wake after the rotor is ignored.

The analysis in section 1.6 can be reworked in terms of the axial interference factor, a, instead of the downstream velocity factor, b. The factor a is defined as the fractional decrease in wind velocity from far upstream to the rotor plane:

$$a = \frac{(V_0 - V_1)}{V_0} \tag{2.4}$$

giving

$$C_p = 4a(1 - a)^2 \tag{2.5}$$

for the power coefficient. The Betz limit is obtained with $a = 1/3$.

Substituting equation (2.4) in equation (1.16) and using equation (1.17), we can show that $b = 1 - 2a$. Hence the velocity induced in the wake after the rotor can be rewritten as $V_2 = V_0(1 - 2a)$. As $V_1 = V_0(1 - a)$ we can see, the reduction in wind speed from far upstream to within the wake is twice that induced at the rotor disc. The range for the axial interference factor a is from 0, for no energy extraction, to $\frac{1}{2}$, where the model implies the wind has slowed to zero in the wake behind the rotor; this is not possible in reality. Outside this range the assumptions made in deriving the simple actuator disc model are violated; in practice the flow behind a real rotor for large values of a becomes highly turbulent and flow reversal eventually occurs, as described by Eggleston and Stoddard (1987).

A more practicable flow model for the rotor is the Glauert *annulus momentum vortex theory*. Each radial section of blade is analysed independently using two-dimensional aerofoil data and equations based on continuity and momentum. Although strictly applicable only to rotors with an infinite number of blades, correction factors can be introduced to allow for the case of a finite number of blades. With a small number of blades the flow is not circumferentially uniform and the annulus theory is used to represent the average affects. This model provides a good approximation to the actual flow and is the most useful model for carrying out basic rotor design; it is described further in section 2.4.

The basic concept of the more advanced *prescribed-wake vortex model* is that the vortices which form a spiral wake behind the rotor blade tips define the flow field around the rotor according to the Biot–Savart theorem. Each section of the rotor blade generates a lift force that is proportional to its local bound vorticity. By integrating the effect of the trailing vortices over the blade, the induced flow and hence the rotor forces and moments can be determined. Aerofoil data are required as in the Glauert momentum vortex model. The path of the trailing vortices has to be defined by an assumed rigid helical wake behind the rotor and a computer is essential for solving the equations.

In the advanced *free-wake vortex model* the assumption of a rigid vortex wake is relaxed and the exact path of the trailing vortices is found iteratively. Needless to say, this approach uses a great deal of computer time and the results differ only slightly from the simpler Glauert momentum model.

2.4 BLADE ELEMENT THEORY

The mathematical details of blade element theory are an optional item

A simple method for calculating the aerodynamic forces on wind turbine blades is known as blade element theory; it is based on the Glauert momentum vortex model. The method is described comprehensively in Eggleston and Stoddard (1987) and in the chapter by Wilson in Spera (1994). It is developed here for a horizontal-axis turbine but can be adapted for vertical-axis turbines.

The rotor blade is divided into radial blade sections (elements) and the airflow from upstream to downstream of the elements is divided into annular stream tubes. It is assumed that each stream tube can be treated

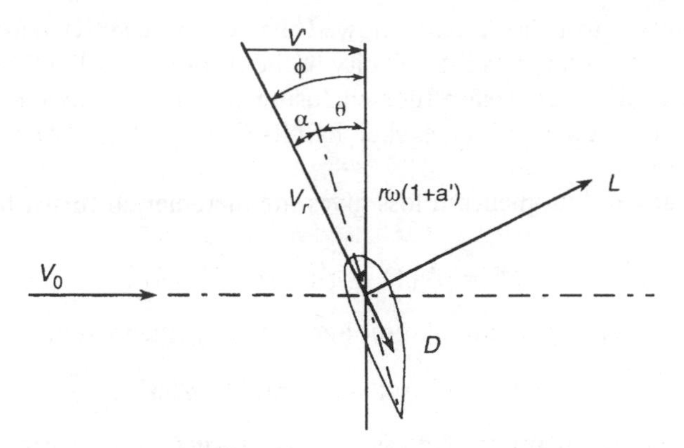

Figure 2.7. Flow around a blade element showing velocity components and flow angles

independently of adjacent ones. The flow around each element is determined, and the aerodynamic forces on each element is calculated using the aerodynamic coefficients. The total load on the blades is calculated by summarising the forces from all the elements.

The definition of one blade element has already been given in Figure 2.5. The complete velocity components and definitions of angles are shown in Figure 2.7.

The relative wind at the rotor V_r varies with radius r and consists of an axial component $V' = V_0(1 - a)$ and a rotational component $r\omega(1 + a')$ where a is the axial interference factor, a' is a tangential interference factor allowing for the induced swirl velocity $w = \omega r a'$ at the rotor disc, and ω is the rotational speed of the rotor. The magnitude and direction of the flow is determined by the rotational speed of the blade element, the wind velocity and the induced axial and swirl velocities.

Knowing the direction and speed of the flow relative to each blade element, the lift and drag are found according to the aerodynamic coefficients for the actual aerofoil. Projection of the forces on the rotational direction gives the driving force, and the projection on the direction parallel to the wind or axial direction gives the thrust force. In terms of the dimensionless lift and drag coefficients, the net torque Q, power W and thrust T caused by B blades of local chord c at each radial blade element are as follows:

$$\mathrm{d}Q = 1/2\rho V_r^2\, r[C_L\sin\phi - C_D\cos\phi]Bc\mathrm{d}r \tag{2.6}$$

$$\mathrm{d}W = 1/2\rho V_r^2\omega r[C_L\sin\phi - C_D\cos\phi]Bc\mathrm{d}r \tag{2.7}$$

$$\mathrm{d}T = 1/2\rho V_r^2[C_L\cos\phi + C_D\sin\phi]Bc\mathrm{d}r \tag{2.8}$$

where

$$V_r = \frac{V'}{\sin\phi} = \frac{(r\omega + w)}{\cos\phi} \tag{2.9}$$

The simple actuator disc theory showed that the induced velocity in the far wake was twice as large as the velocity in the rotor plane. This is assumed to hold for each annular stream tube enclosing each radial blade element. It is also assumed that the air has reached half its final rotational or swirl velocity at the rotor disc.

The linear axial momentum loss gives the incremental thrust force on the rotor as

$$dT = \rho V_0(1-a)(2\pi r dr)2(aV_0) \tag{2.10}$$

and the rate of change of angular momentum gives the incremental torque as

$$dQ = \rho V_0 r(1-a)(2\pi r dr)2(a'\omega r) \tag{2.11}$$

Equating the momentum expressions for thrust and torque with their counterparts obtained using aerodynamic forces on a blade element, we obtain the final equations:

$$\frac{a}{1-a} = \left(\frac{\sigma R}{8r}\right)\left(\frac{C_L\cos\phi + C_D\sin\phi}{\sin^2\phi}\right) \tag{2.12}$$

$$\frac{a'}{1+a'} = \left(\frac{\sigma R}{8r}\right)\left(\frac{C_L\sin\phi - C_D\cos\phi}{\sin\phi\cos\phi}\right) \tag{2.13}$$

$$\tan\phi = \frac{V_0(1-a)}{\omega r(1+a')} = \frac{R}{rX}\left(\frac{1-a}{1+a'}\right) \tag{2.14}$$

where the blade solidity (ratio of blade area to total rotor swept area) is $\sigma = Bc/\pi R$ and the tip speed ratio is defined by $X = \omega R/V_o$.

The two-dimensional lift and drag coefficients are both functions of the angle of incidence α where $\alpha = \phi - \theta$, and can be found from tables if a standard aerofoil is used. Equations (2.12) to (2.14) can be solved iteratively for a and a' for any pitch angle θ, assuming convergence can be obtained; there will not be convergence when a approaches 0.5. The solution procedure is to guess values for a and a'; ϕ is calculated from equation (2.14) then $\alpha = \phi - \theta$ from which values of the lift and drag coefficients can be found. The interference factors a and a' can then be updated from equations (2.12) and (2.13) and the cycle repeated until convergence is achieved. The same procedure is repeated at different radial positions along the blade, and the overall power and thrust of the rotor can be found by summing the values for each blade element:

$$W = \sum_{i=1}^{i=n} \omega dQ \tag{2.15}$$

$$T = \sum_{i=1}^{i=n} dT \tag{2.16}$$

where n is the number of blade elements.

Accuracy can be increased by applying various correction factors, such as the tip loss factor. The tip loss factor allows for the velocities and forces not being circumferentially uniform due to the rotor having a finite number of blades. The Prandtl tip loss factor can be expressed as

$$F = (2/\pi)\ \arccos\left(\exp\left\{0.5\left[1 - \frac{r}{R}\right]B\sqrt{1 + X^2}\right\}\right) \qquad (2.17)$$

and is applied in the form $V' = V_o(1 - aF)$ for the axial induced velocity at the rotor.

In recent advanced aerodynamic computational methods the lift and drag coefficients can be calculated directly for each blade element instead of interpolated from tables for standard aerofoils. However, in the stall region, at high angles of incidence, the lift and drag coefficients usually have to be estimated from empirical correlations. For $a > 0.45$ a correlation based on the observed behaviour of the thrust coefficient is used to overcome the breakdown in the theory that occurs.

For optimum performance the blade must be twisted with a large pitch angle near the root, gradually decreasing outwards so the tip section is nearly tangential. This arises from the need to keep the angle of incidence of the flow onto the blade below the stalling angle as the tangential speed of the blade increases with radius. For structural reasons the drag or thrust force on the rotor should be kept to a minimum; this reduces the bending stresses at the blade roots.

Notice how this calculation procedure requires initial values for the blade solidity and the tip speed ratio. This in turn means that the plan-form or radial variation in the blade chord has to be known, as well as the rotational speed and wind speed. Selection of these parameters forms an important part of designing a new wind turbine rotor and depends mainly on the specific application. The rotor is optimised overall by optimising each blade element to obtain the maximum energy extraction; various optimising strategies are described by Eggleston and Stoddard (1987).

SUMMARY OF THE CHAPTER

This chapter has described basic aerodynamic ideas relating to wind turbine design. The different classes of wind turbine were outlined, including lift and drag type rotors, horizontal and vertical wind axis machines and fixed and variable speed designs. By treating rotor blades as aerofoils it was shown how standard aerofoil theory can be used to calculate the forces on the blades and hence determine the power produced by the rotor. The annulus momentum theory, based on dividing the rotor blades into radial elements, was described in some detail as it is often used in preliminary rotor analysis.

BIBLIOGRAPHY AND REFERENCES

Abbott, I.H. and Von Doenhoff, A.E., *Theory of Wing Sections*, Dover, New York, 1959.

Eggleston, D.M. and Stoddard, F.S., *Wind Turbine Engineering Design*, Van Nostrand, New York, 1987.
Freris, L.L. (ed.), *Wind Energy Conversion Systems*, Prentice Hall, London, 1990.
Glauert, H., *The Elements of Aerofoil and Airscrew Theory*, 2nd ed., Cambridge University Press, Cambridge 1959.
Lipman, N.H., Musgrove, P.J. and Pontin, G.W.W. (eds), *Wind Energy for the Eighties*, British Wind Energy Association, Peter Peregrinus, Stevenage, UK, 1982.
Spera, D.A. (ed), *Wind Turbine Technology, Fundamental Concepts of Wind Turbine Engineering*, ASME Press, New York, 1994.
Taylor, R.H., *Alternative Energy Sources for the Centralised Generation of Electricity*, Adam Hilger, Bristol, 1983.
World Meteorological Organisation, *Meteorological Aspects of the Utilisation of Wind as an Energy Source*, Technical Note 175, WMO, Geneva, 1981.

SELF-ASSESSMENT QUESTIONS

PART A. True or False?

1. Modern wind turbines for electricity production have many blades.

2. The rotor of a vertical-axis turbine experiences cyclic stresses during each revolution but they are not due to gravitational forces.

3. The simple actuator disc model takes into account the rotation of the wake.

4. There are three forces acting on a two-dimensional aerofoil when moving in a fluid.

5. As the angle of incidence increases for a two-dimensional aerofoil the flow first starts to separate on the suction surface.

6. Blade element theory cannot be used to analyse vertical-axis wind turbines.

7. The annulus momentum theory (a) is based on three-dimensional fluid flow (b) assumes a finite number of blades and (c) can take account of non-uniform flow between the blades.

8. In the prescribed-wake vortex model the path of the trailing vortices from the rotor blades is determined iteratively using a computer.

9. In the momentum and blade element analysis of turbine rotors (a) the air velocities at the rotor plane are based on the induced velocities predicted by the actuator disc theory, (b) the full equations have to be solved by iteration, and (c) the equations can be solved for all values of the induced velocity factors.

10. The radial variation in the angle of incidence of the flow onto the rotor is independent of the angular speed of the rotor.

PART B

1. What is the ratio of the efficiency of a propeller turbine to that of a Savonius rotor?

2. Contrast the main characteristics of low and high speed wind turbines.

3. In wind turbine aerodynamic theory what is the significance of assuming an infinite number of blades?

4. How do the lift and drag coefficients for a two-dimensional aerofoil vary with incidence angle?

5. What is the basic assumption made in blade element theory?

6. How does the angle of incidence vary along a rotor blade if it is not twisted?

7. A non-rotating untwisted rotor blade of length 10 m and mean chord 0.5 m is placed in an airflow at standard conditions; the airflow has a velocity of 20 m/s. Experiments show that before separation the lift coefficient from zero incidence increases by 0.1 per degree of incidence and the drag coefficient remains almost constant at 0.004. What are the lift and drag forces at an angle of incidence of 10°? If the forces act at the quarter-chord point, what is the pitching moment about the nose of the aerofoil?

8. Derive one expression for the blade solidity of a rotor.

9. How can non-uniform circumferential flow effects be accounted for in the blade element momentum theory?

10. Which force causes useful torque in the direction of the rotation of the blades?

Answers

Part A

1. False; 2. True; 3. False; 4. True; 5. True; 6. False; 7. (a) and (b) False (c) True; 8. True; 9. (a) and (b) True (c) False; 10. False.

Part B

1. The C_p for a propeller turbine is about 0.42 and for a Savonius rotor about 0.15, so the ratio is 2.8.

2. Low speed turbines have many blades and hence high solidity, high starting torque and low rotational speed. High speed turbines have three blades or less, thus they have low solidity, low torque and high rotational speed.

3. An infinite number of blades implies there are no circumferential flow variations.

4. See Figure 2.4.

5. Assume that the flow around each individual blade element can be analysed independently from adjacent elements.

6. The blade speed and hence the relative tangential air velocity increases with radius, so for a constant wind velocity the angle of incidence decreases from root to tip.

7. 1200 N and 4.8 N, 147.82 N m anticlockwise if the flow is left to right.

8. Blade solidity,
$$\sigma = \frac{\text{total blade area}}{\text{swept rotor area}} = \frac{B(cR)}{\pi R^2} = \frac{Bc}{\pi R}$$

   This is a simple and commonly used definition but others are also used.

9. By use of correction factors applied to the induced velocity factors, such as the Prandtl tip loss correction factor.

10. The lift force causes the blades to rotate and generate power; the drag force opposes this motion.

3

Components and Operational Characteristics

AIMS

This unit describes the layout and main components of a typical horizontal-axis wind turbine for electricity generation, and the key concepts relating to the operational characteristics of a single wind turbine and of wind farms when machines are grouped together.

OBJECTIVES

At the end of this unit you will be able to do four things :

1. Call upon a basic knowledge of the construction and operation of horizontal axis wind turbines for electricity production.
2. Describe the function of the main mechanical, structural, control and electrical components of a horizontal-axis wind turbine.
3. Understand the meaning of the key parameters related to the operation of a wind turbine or wind farm.
4. Calculate the annual output of a wind turbine or wind farm from the key parameters.

3.1 GENERAL LAYOUT

The general layout of a typical horizontal-axis wind turbine is shown in Figure 3.1. The energy is extracted from the wind by the rotor and is either used directly as mechanical energy, such as in a water pump, or converted to electrical energy by a generator.

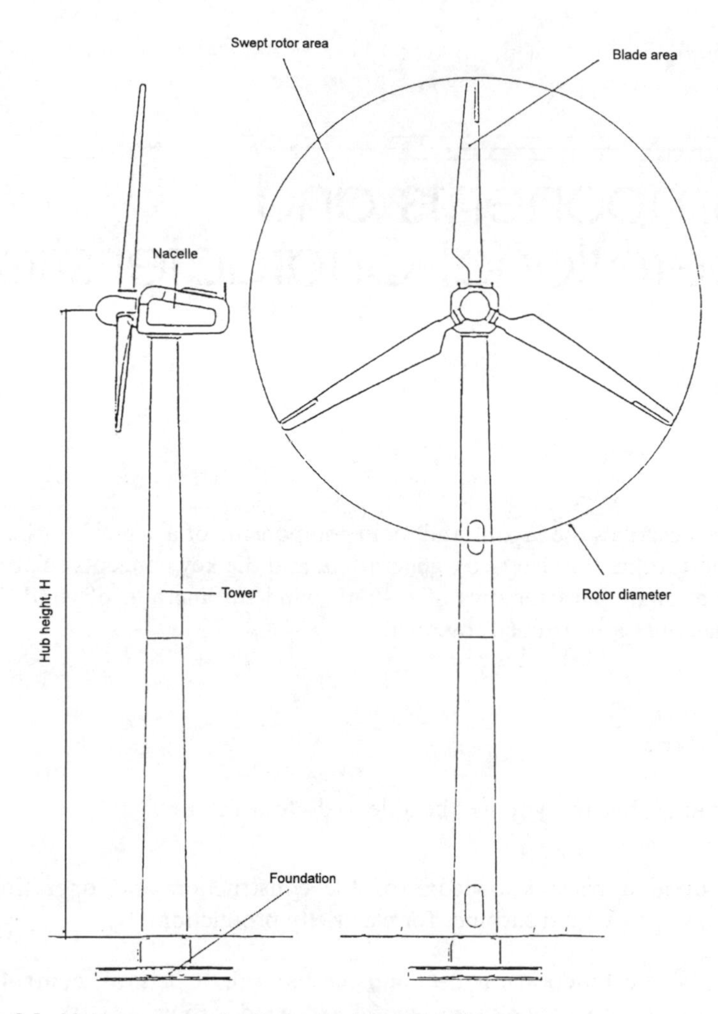

Figure 3.1. Main components of a horizontal-axis wind turbine

The following parameters are normally used to specify a wind turbine:

- *Hub height*: the height of the hub above ground level
- *Swept area*: the area defined by the rotating rotor disc
- *Solidity*: the ratio of the area of the blades to the swept area
- *Tip speed ratio*: the ratio of the speed of the blade tip to the wind speed
- *Rated power*: Maximum continuous power output at the electrical connection point.

The following section describes the main components of a grid-connected wind turbine. When used on a wind farm their rated power may be 200–750 kW and their rotor diameters may be 25–50 m. Commercial prototypes of up to 1.5 MW are under test in Europe and their characteristics are described by Hau *et al.* (1993).

Very large turbines of up to 4 MW and 100 m rotor diameter have been built in the past as experimental machines but their performance was disappointing. It is likely that the diameter and power rating of commercial wind turbines will continue to increase slowly with time until an economic optimum is reached. At present, the rotor diameter that gives the most cost-effective performance remains a subject of debate.

3.2 MAIN DESIGN FEATURES

The main components of a wind turbine for electricity generation are the rotor, the transmission system, the generator, and the yaw and control systems; their layout is shown in Figures 3.2 and 3.3. The major components are fixed onto or inside a nacelle, which can rotate (or yaw) according to the wind direction and which is mounted on the tower.

Electricity systems function within close limits on operating parameters such as voltage, frequency and harmonic content. On large national electricity systems, wind turbines have effectively no impact on the system

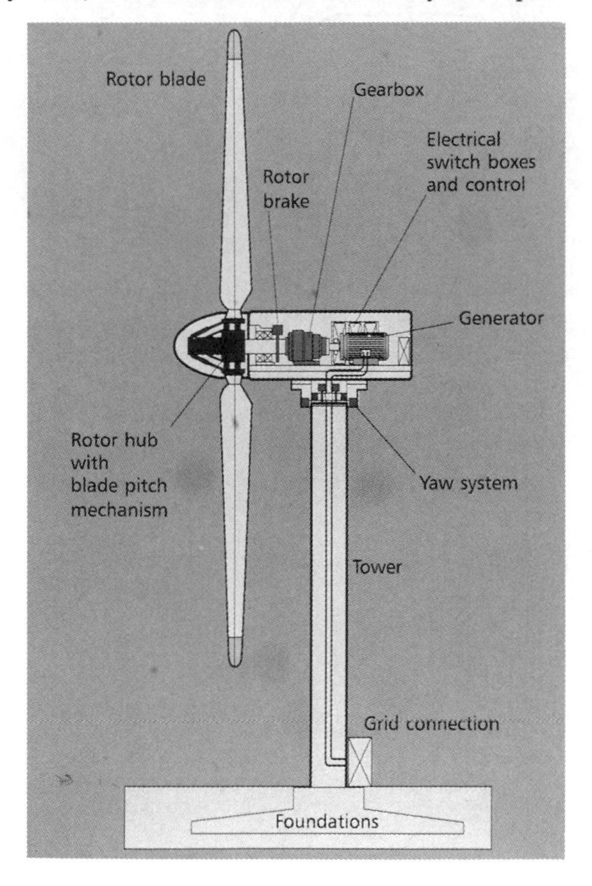

Figure 3.2. Cross-section of a typical grid-connected wind turbine

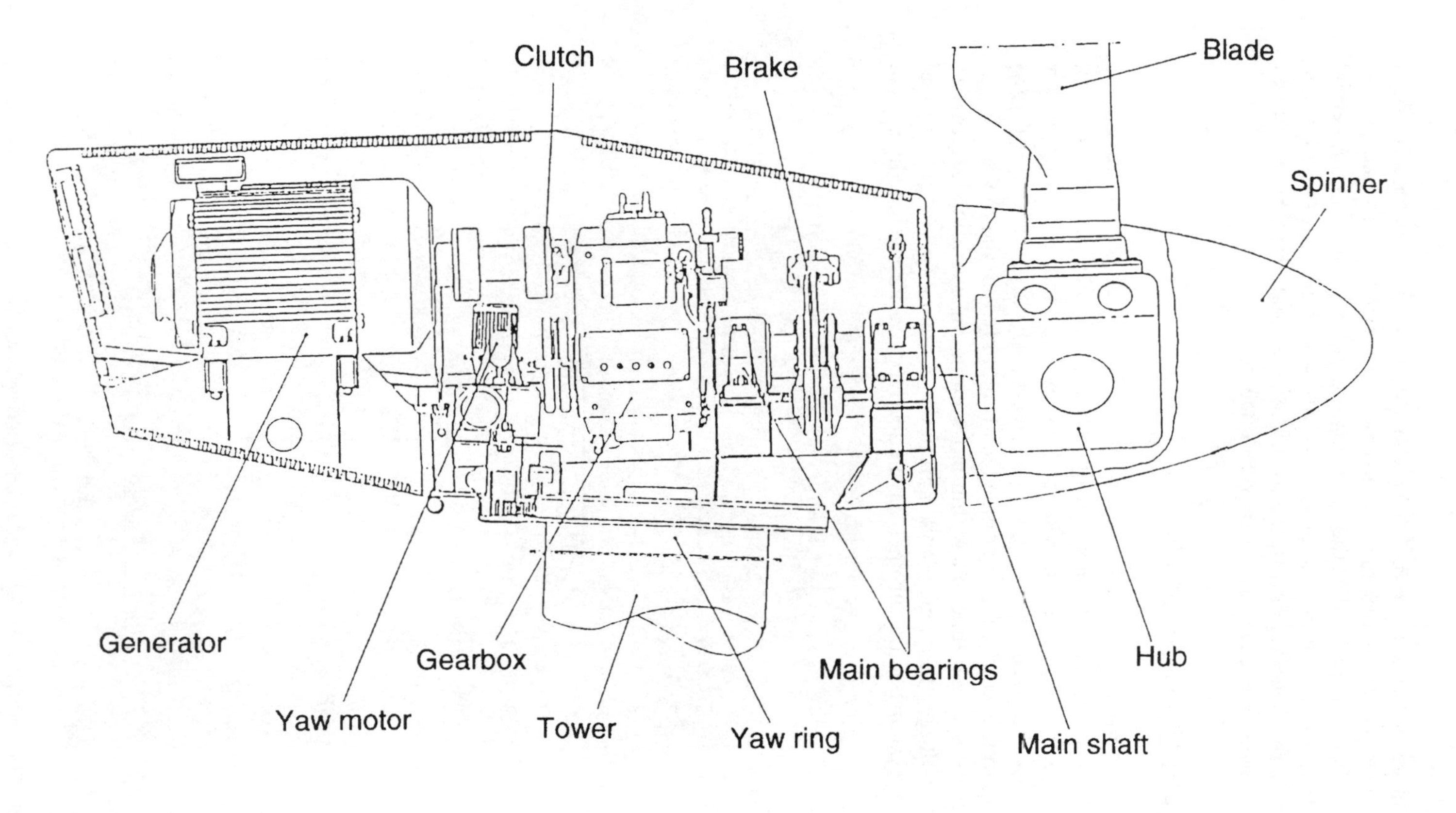

Figure 3.3. Cross-section of a nacelle in a grid-connected wind turbine

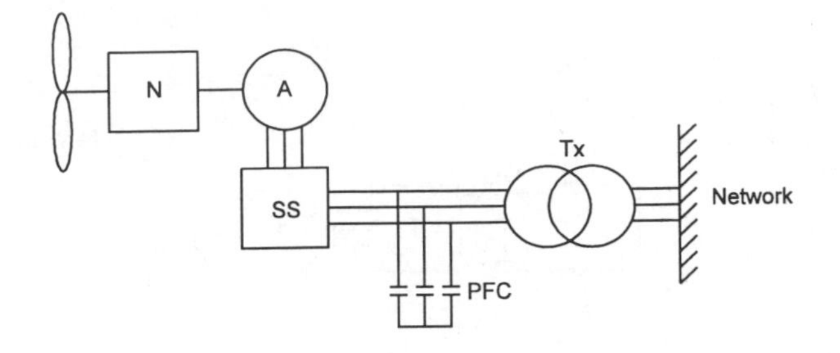

Figure 3.4. Schematic of a fixed-speed wind turbine generator: N = gearbox, A = asynchronous or induction generator, SS = soft-start, PFC = power factor correction, Tx = transformer

frequency and usually they have only a limited effect on local network voltage. Wind turbines are generally used as a source of energy rather than as control devices to maintain the quality of power on the utility network. In fact, for many installations, careful calculations are required to ensure that the wind turbines do not degrade the power quality of the local utility system.

The most common arrangement is for the generator to be connected direct to the utility system. This is shown schematically in Figure 3.4.

A three-phase asynchronous (induction) generator is used, usually generating at 690 V. At the base of each tower a transformer increases the voltage to, say, 10 kV for transmission to the utility network. Because of the characteristics of the induction generator a soft-start unit is used to connect the generator to the network and capacitors are installed to improve the power factor. The soft-start and capacitors are normally located in a cabinet in the base of the tower together with the switchgear and controller equipment.

An induction generator operates at a speed that is effectively constant, fixed by the frequency of the utility system, so the wind turbine rotor must also operate at a fixed speed as the gearbox has a constant ratio. As discussed in Chapter 2, this means the wind turbine cannot operate at maximum efficiency over a wide range of wind speeds. Some early designs attempted to overcome this limitation by using two generators with different operating speeds, both driven from the high speed shaft of the gearbox. The smaller, slow speed generator was used in low wind speeds. As the wind speed increased, the slow speed generator was disconnected and the larger, high speed generator brought into service. In some modern large wind turbines this two-speed concept has been developed by placing two different speed windings inside the same generator. Two-speed operation can significantly increase the number of hours a turbine is generating but may have only a modest effect on annual energy yield. However, a major advantage is that, in low winds, the tip speed of the blades is reduced, reducing the aerodynamic noise.

Recent developments in power electronics have made continuously variable-speed drives a practical possibility and this concept can be cost-effective in some circumstances. Figure 3.5 shows one approach to variable-speed operation but, as this technology is still evolving, there are several

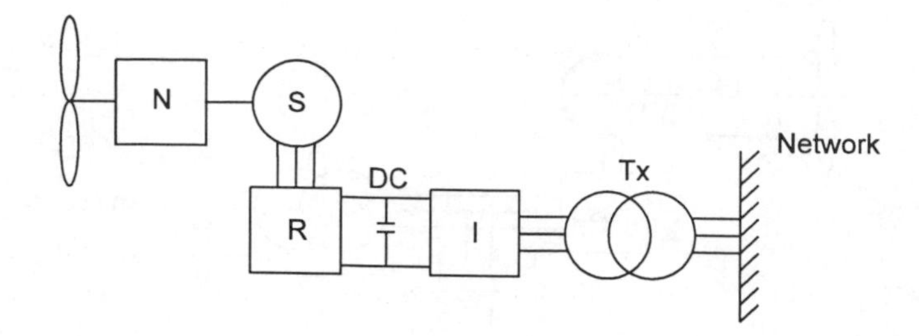

Figure 3.5. Schematic of a variable-speed wind turbine generator: N = gearbox, S = synchronous generator, R = rectifier, I = inverter, DC = direct current link, Tx = transformer

alternative architectures. In the arrangement of Figure 3.5 a conventional rotor and gearbox are coupled to a synchronous generator. A rectifier unit at the tower base converts the three-phase alternating current (AC) to direct current (DC). The DC bus-bar is very short and, adjacent to the rectifier, is an inverter, which converts the DC to AC at the 50 Hz of the utility network. The fundamental principle of operation is that, because all the power is rectified to DC then inverted to AC, the generator is now decoupled from the network and can operate over a wide range of speeds. Early variable-speed drives used thyristors as switching elements and this generated significant harmonic currents. Contemporary designs use transistors as switches and, as they can be operated much faster, the waveforms generated are much closer to sine waves with lower harmonic distortion.

3.2.1 Rotor

Turbines for wind farm applications typically have two or three blades and a tip speed of 50 to 70 m/s. With these tip speeds a 3-bladed rotor usually gives the best efficiency, though 2-bladed rotors are only 2-3% less efficient. It is even possible to use a single-bladed rotor with a counterweight for balance with a further small drop in efficiency, typically 6% below that of 2-bladed rotors. Although with less blades there are of course fewer blades to pay for, there are penalties. For similarly proportioned blades, rotors with less blades have to run faster and this can lead to problems of tip noise and erosion. Three blades are also considered by many to be aesthetically most pleasing. The forces are more evenly balanced in a 3-bladed rotor and the hub can be simple whereas 2 and 1-bladed rotor hubs often have the complication of teetering to limit fluctuating forces as the blades sweep the wind through a varying velocity field. With a teetered hub, the rotor is hinged at the hub in such a way as to allow the plane of rotation of the rotor to tilt backwards and forwards through a few degrees away from the mean plane. The rocking motion or teetering of the rotor which results at each revolution significantly reduces the loads on the blades due to gusts and wind shear.

Blades have been constructed from glass reinforced plastic (GRP), wood and wood laminate, carbon fibre reinforced plastic (CFRP), steel and aluminium. For small wind turbines, say less than 5 m diameter, the issue of the material is usually driven by production efficiency rather than the weight, stiffness or other particular characteristics of the blade. For larger wind turbines, the blade characteristics are usually more difficult to obtain and are more influential over material choice.

Most larger wind turbine blades in the world are made from GRP, and most of these by hand lay up with polyester resins as is used commonly for boat hulls, garden, play equipment and a range of consumer goods world-wide. The process does not require high levels of skill to obtain a reasonable result and if one is not too concerned about weight, say for smaller blades up to 20 m in length, neither is the design challenging. There are, however, many much more sophisticated ways to use GRP which reduce the weight and increase stiffness which cannot be covered in detail here. The glass is laid more accurately for example by applying it in pre-impregnated sheets, a higher performance resin is used such as epoxy in more controlled proportions often vacuum-bagged and cured at raised temperatures. Currently, it does appear that the simpler hand laid polyester, with some care in the selection and placement of fibres, offers the lowest cost solution for large GRP blades.

Wood has a lot to commend it having been used for the earliest historical machines. It has low weight, cost and excellent fatigue strength, the drawbacks being sensitivity to moisture and processing costs. A technique called 'cold-moulding' is, however, being employed which overcomes these problems. Wood veneers are laminated with epoxy resin under a vacuum bag which presses them to the shape of the blade mould. The blade is otherwise similar to a GRP blade with skins of glass fibre. The wood, however, confers greater stiffness with less overall weight which becomes increasingly important at larger sizes. CFRP blades have been made successfully in prototype and limited production runs. It confers the highest stiffness and lowest weight but unfortunately is expensive material. There has been a hope that carbon fibre will become cheaper as demand increases but currently the price is rising steadily and it will not have widespread use for blades until it becomes much cheaper.

Steel was thought to be a strong contender for blade materials some years ago. Unfortunately the weakness with steel in this application is the relatively low fatigue strength and the high density which leads to high self-weight induced alternating stresses. Aluminium had been used only for experimental blades and even than with only partial success since the fatigue strength is worse than that of steel.

Blades made from the non-metallic materials all share a difficult design detail at the root where the bending moments are greatest and the change in stiffness between the blade material to the steel hub leads inevitably to stress concentrations. Established manufacturers have now tested and proved their own proprietary solutions in a variety of ways including massive bonded-on metal tubes and embedded steel studs.

It has been said that the best conceivable wind turbine blade aerodynamically would be only 10% more efficient than a plank of wood! This is partly true because the range of wind directions and speeds which a blade has

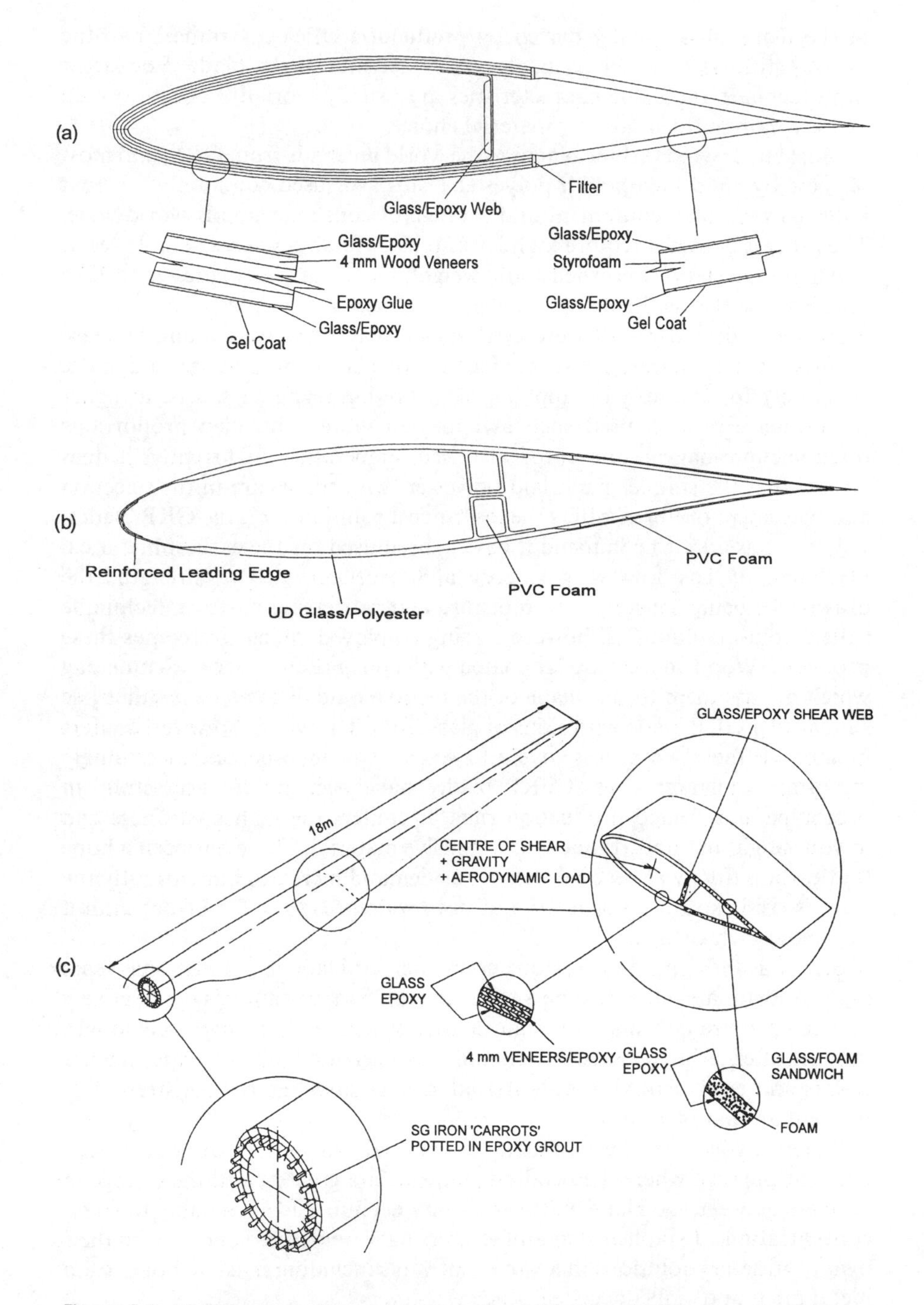

Figure 3.6. (a) Typical wood/epoxy blade section (b) Typical GRP blade section (c) Cross section of wood/epoxy composite blade. Reproduced by permission of Taywood Aerolaminates.

to cope with tends to blur the distinctions. However a well chosen aerofoil will have a number of characteristics which if not all essential will improve the machine greatly:

- high lift-to-drag (L/D) for efficiency over a wide range
- good stall characteristic
- insensitivity to roughness
- low noise production

In addition the shape should be compatible with the production process although with moulded blades this is not a major limitation. Also, the aerofoil should allow enough room for the structure within to take the loads it develops.

In recent years, blades employing new aerofoils have often under-performed and with the investment in new blade development and tooling being so high, designers tend today to be very conservative using well-proven aerofoils with publicly-available data such as NACA 632XX , NACA 634XX and NASA LS-104XX. Blades tend to have slightly higher lift aerofoils closer to the root which gradually transform to lower lift aerofoils near the tip.

This leads on to the general shape of the blade. Modern blades tend to have a circular root which gradually changes to a thick aerofoil at 20 to 30% radius at the largest chord. The aerodynamic part which is outboard of this becomes thinner towards the tip, tapers in chord and twists. The taper, twist and aerofoil characteristics should all combine to give the best possible energy capture for the rotor speed power and site conditions. Programmes are available on the market to predict performance. Stall performance is however still poorly understood and predictions are usually empirical/based on monitored data from a wind turbine. The uncertainty in stall prediction is another reason why aerofoil choice remains limited.

The rotor may be mounted on the upwind side of the tower or downwind. In the latter case, the blades can be coned outwards to increase tower clearance. A downwind rotor will, in principle, align itself with the wind direction so removing the need for a yaw drive system. However, downwind rotors suffer increased noise and cyclic loads due to the tower "shadow" and are less common than upwind designs.

As discussed in chapter 2, the output of the rotor can be regulated in a number of ways:

For *stall regulated* turbines, the pitch angle distribution along the blades is constant for all wind speeds. At high wind speeds, the angle of incidence on to the blades increases. The lift forces are reduced duc to stall, but thc drag forccs will rise and the increase in power which would otherwise occur is prevented. This results in relatively large aerodynamic forces on the rotor. Stall regulation is, however, very simple and does not require any control system. There remains considerable difficulty in designing stall regulated rotors and research is continuing on both 3-dimensional and dynamic stall effects.

Figure 3.7. Testing a wind turbine blade. Reproduced by permission of Taywood Aerolaminates

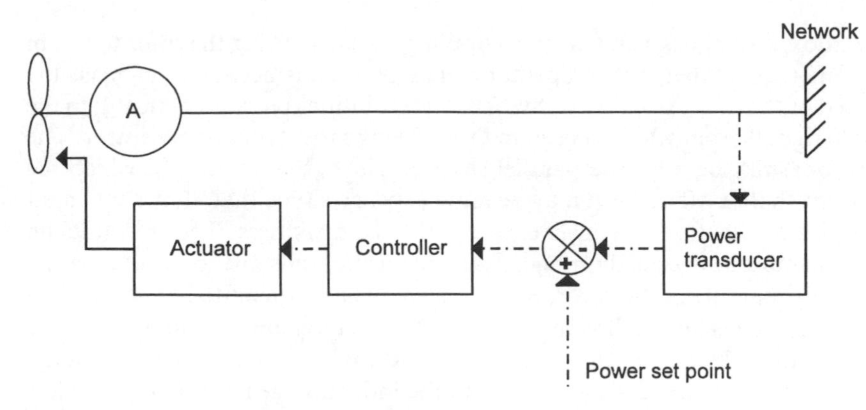

Figure 3.8. Pitch regulation control system

For a *pitch-regulated* turbine the blades can be rotated about their radial axis during operation as the wind velocity changes. It is therefore possible to have an almost optimum pitch angle at all wind speeds and a relatively low cut-in wind velocity. At high wind speeds, the pitch angle is changed in order to reduce the angle of attack and hence the aerodynamic driving forces on the blades. This ensures that the power output from the rotor is limited to the rated power of the generator. In some designs only the outer part of the blades is movable. The aerodynamic load on the rotor is reduced compared to the load on a stall-regulated rotor. Pitch regulation is more expensive and requires a relatively complicated control system but is inherently more efficient than stall regulation. Figure 3.8 shows the outline of the conventional control system used for a pitch-regulated turbine.

The power generated by the turbine is measured using an electrical power transducer, normally situated in the base of the tower. The measured signal is compared with the set point (the rated power of the turbine) and the error signal is then passed to the controller. Here a control algorithm is used to obtain the required dynamic response for the actuator to rotate the blades. Several types of actuator have been used, including hydraulic and electromechanical devices. The difficulty is to change the pitch of the blades on the constantly rotating rotor. The feedback system operates continuously to regulate the power above rated wind speed, but there are a some practical difficulties. In turbulent winds it is not uncommon to encounter significant transient power excursions well above the set point.

A further system used on a few wind turbines is *yaw control.* In this arrangement the entire rotor is rotated horizontally or yawed out of the wind. Although a passive form of yaw control is used on wind pumps, it is rare in electricity-generating turbines.

3.2.2 Transmission System

The mechanical power generated by the rotor blades is transmitted to the generator by a transmission system located in the nacelle. This consists of a gearbox, sometimes a clutch, and a braking system to bring the rotor to rest in an emergency when not in operation. The gearbox is needed to increase the speed of the rotor, from typically 20 to 50 revolutions per minute (rpm), to the 1000 or 1500 rpm which is required for driving most types of generator. The gearbox may be a simple parallel-shaft gearbox (Figure 3.3), in which the output shaft is offset, or it may be more expensive type that allows the input and output shafts to be in line, for greater compactness. The transmission system must be designed for high dynamic torque loads due to the fluctuating power output from the rotor. Some designers have attempted to control the dynamic loads by adding mechanical compliance and damping into the drivetrain. This is particularly important for very large wind turbines, where the dynamic loads are high and the induction generator provides less damping than at smaller sizes.

3.2.3 Generator

All grid-connected wind turbines drive three phase alternating current (AC) generators to convert mechanical to electrical power. Generators are divided into two main classes. A synchronous generator operates at exactly the same frequency as the network to which it is connected; synchronous generators are also called alternators. An asynchronous generator operates at a slightly higher frequency than the network; asynchronous generators are often called induction generators.

Induction generators and synchronous generators each have a non-rotating part called the stator. The stators are similiar for the two types of generator. Both types of stator are connected to the network and both consist of a three-phase winding on a laminated iron core; they produce a magnetic field rotating at a constant speed. But though the generators have similar stators, their rotors are quite different. In a synchronous generator the rotor has a winding through which there passes a direct current; this is the field winding. The field winding creates a constant magnetic field which locks into the rotating field created by the stator winding. Thus the rotor always rotates

Real, reactive and apparent power

Electrical power engineers think in terms of real, reactive and apparent power flows in a power system. Real power (kW or MW) is the capacity to do useful work such as pump water or rotate a shaft against a load. Reactive power (kVAr or MVAr) is drawn by transformers or induction motors as magnetising current for their iron cores but is generated by capacitors. Finally, apparent or complex power (kVA or MVA) is the vector addition of real and reactive power. The power triangle of an inductive load is shown in Figure 3.9. Here the load takes real power, P, to provide heat or work and some reactive power, Q, to energise the magnetic circuits. The power factor is defined as the ratio ρ/S or cos ø.

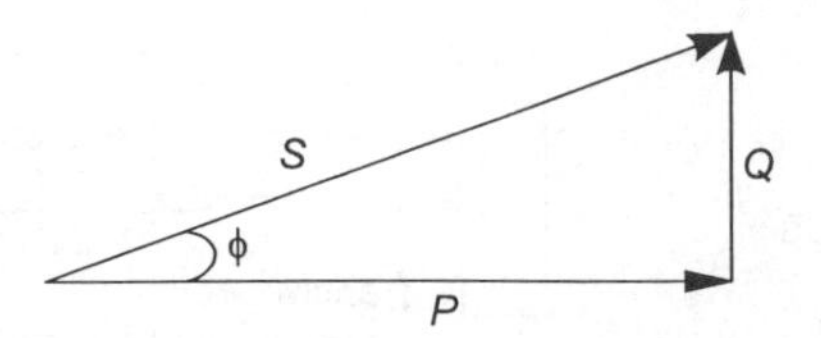

Figure 3.9 Power triangle of an inductive load: P = real power (W), Q = reactive power VAr, S = apparent complex power (VA)

at a constant speed *in synchronism* with the stator field and the network frequency. In some designs the rotor magnetic field is produced by permanent magnets, but this is not common for large generators.

The rotor of an induction machine is quite different. It consists of a squirrel cage of bars, short-circuited at each end. There is no electrical connection onto the rotor, and the rotor currents are *induced* by the relative motion of the rotor against the rotating field of the stator. If the rotor speed is exactly equal to the speed of the rotating field produced by the stator(as in the synchronous machine), there is no relative motion, so there is no induced rotor current. Therefore an induction generator always operates at a speed that is slightly higher than the speed of the rotating field stator. This difference in speed is known as the slip and is likely to be approximately 1% during normal operation. Induction generators are not common but there are millions of induction motors in service throughout the world. An induction generator is essentially an induction motor with torque applied to the shaft rather than taken from it.

There are many standard textbooks describing synchronous generators and induction motors and generators. McPherson (1981) gives a particularly clear description of how an induction generator operates.

In an electrical power system almost all the large generators are synchronous generators. They are slightly more efficient than induction generators and have the major advantage that their reactive power flow can be controlled. In a synchronous generator the direct current flowing in the field winding magnetises the rotor, so by increasing the field current, reactive power may be exported to the network. Similarly, by reducing the field current, reactive power may be drawn from the network. Controlling the flow of reactive power gives control over the voltage of the power system.

Some of the early prototype wind turbines were fitted with synchronous generators. However, a synchronous generator is 'locked' into the frequency of the network and this connection can be thought of as a large spring (Figure 3.10). Now a horizontal-axis wind turbine rotor produces torque pulsations at the frequency with which the blades pass the tower. If these torque pulsations are at the natural frequency of the spring–mass system created by the blades and their network connection via the synchronous generator, there will be resonance and some very large oscillations of the drive-train. This is precisely what occurred with one early wind turbine and, although it is possible to provide damping in the drivetrain with mechanical devices such as fluid couplings, nowadays synchronous generators are not normally used in fixed-speed wind turbines.

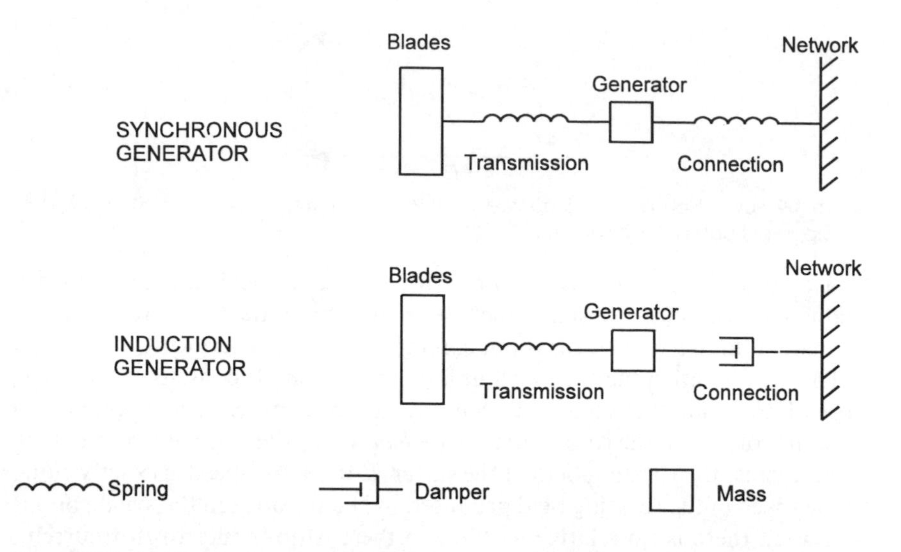

Figure 3.10. Mechanical analogues of synchronous and induction generators

In contrast an induction generator is not locked to the frequency of the network, instead it slips by operating at a slightly higher speed. Therefore the cyclic torque fluctuations at the wind turbine rotor can be absorbed by very small changes in the slip speed. The connection of an induction machine to the network can then be thought of as a damper rather than a spring.

The performance of an induction machine can be described by its torque–slip curve. A torque–slip curve for the generator used on a 600 kW fixed-speed wind turbine is shown in Figure 3.11. Slip is shown along the x-axis; 1 corresponds to standstill and zero corresponds to rotation in synchronism with the stator field. The convention of declaring the slip to be positive for speeds below synchronous comes from the very widespread use of induction motors. Torque is shown on the Y-axis with 1 per unit torque corresponding to the rating of 600 kW.

The curve illustrates how the same machine may operate as a motor or generator. Motor operation is shown between slips of 1 and 0. The normal range of operation of a generator is between the points marked O and A. At point O there is no torque applied to the shaft and, as the wind turbine rotor and transmission apply torque to the generator, the operating point moves towards A. At point A the generator will be producing 600 kW at a speed slightly above synchronous.

Unfortunately there are several disadvantages in using induction generators. Unlike a synchronous generator, which can operate at almost any power factor, the induction generator draws reactive power depending on the real power output. Figure 3.12 shows part of the so-called circle diagram of an induction generator.

At no real power output, point O in Figure 3.11, the generator still draws considerable reactive power to magnetise its iron core. As torque is applied and the operating point moves towards A, real power is exported to the

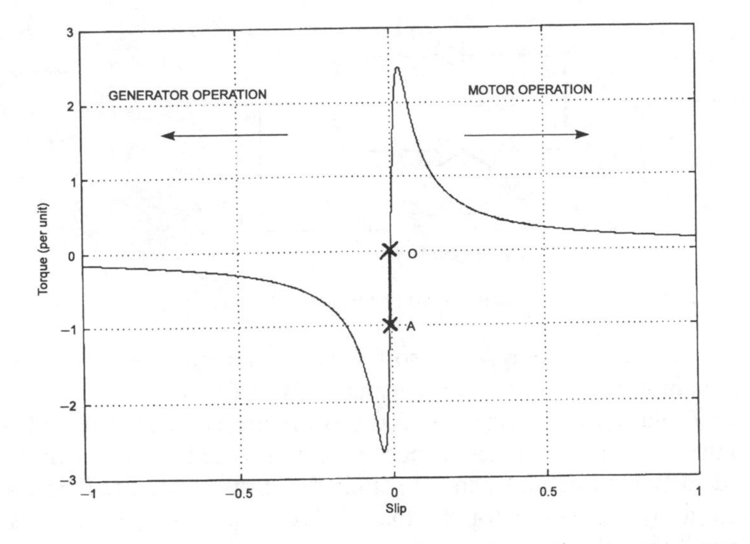

Figure 3.11. Torque-slip curve for an induction machine

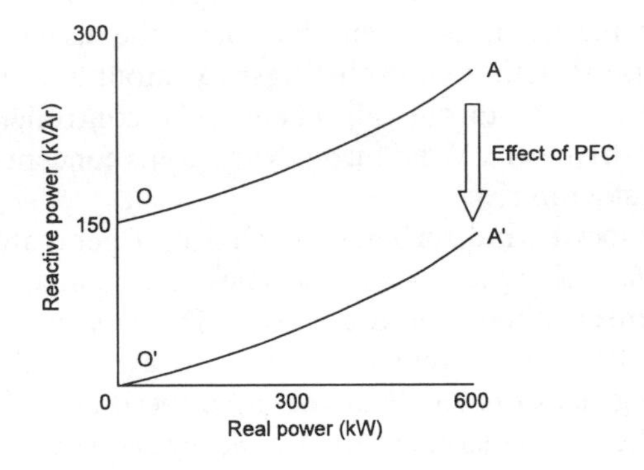

Figure 3.12 Operating characteristic of an induction generator

network but more reactive power is absorbed. Absorbing reactive power is often undesirable as it leads to higher electrical losses in the network. Therefore, as shown in Figure 3.4, capacitors are connected at the base of the wind turbine tower to provide reactive power or *power factor correction (PFC)*. It is usual to provide PFC to compensate for the reactive power demand of the generator at zero output (point O′ in Figure 3.12). At full output (point A′) there is still some reactive power demand. It is certainly possible to connect even more capacitors, but this can lead to hazardous overvoltages in a resonant condition, known as self-excitation, if the connection to the network is lost.

An additional difficulty with using induction generators is that, when they are connected to the network, there is a very large inrush of current as the iron core is magnetised and the normal operating conditions established. This is similar to the direct-on-line starting problem of induction motors. Wind

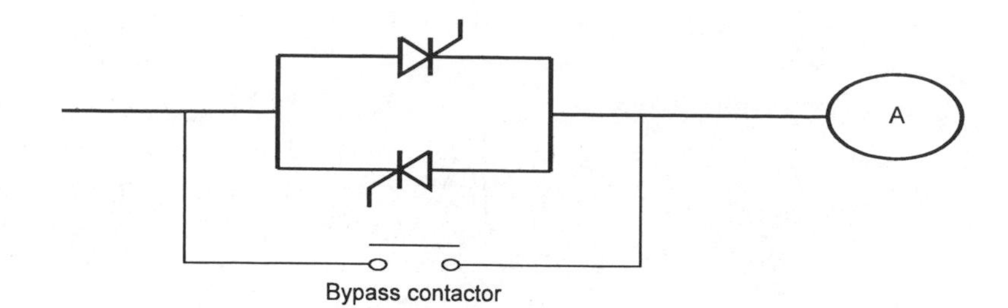

Figure 3.13. Wind turbine soft-start (single phase shown)

turbines use the same type of soft-start equipment used to start large induction motors. This is shown schematically in Figure 3.13.

In each phase of the supply two thyristors are connected back-to-back. When the induction machine is connected they are used to control the voltage applied to the stator and hence to limit the inrush current. It is usual to connect a bypass contactor so that, as soon as the generator is fully connected, the soft-start is switched out.

The use of induction generators in fixed-speed wind turbines is now well established with most manufacturers following a similar approach. However, one interesting recent development has been the use of a wound-rotor induction generator with a controlled resistor mounted on the rotor. This allows the shape of the torque–slip curve to be controlled, improving the dynamics of the drivetrain. Wind turbines using this concept are being sold by one major manufacturer.

For variable-speed wind turbines the choice of generator is much more open. Some variable-speed wind turbines use conventional induction generators whereas others use alternators. There is increasing interest in large-diameter direct-drive generators that do away with the need for a gearbox. These generators rotate at the same speed as the blades, but their design allows them to generate at the frequency required. There are several successful direct-drive wind turbines in service using large-diameter wound-field synchronous generators; the use of direct-drive, permanent-magnet synchronous generators is also under development.

There are also rapid developments in the converters used in variable-speed wind turbines. Modern designs use rapidly switched transistors to produce output voltage waveforms very close to sine waves. This technique creates only small harmonic distortions of the network voltage and it allows the power factor of the wind turbine's the output to be controlled. However, fast switching of semiconductor devices can cause significant power losses and heat generation.

The generator normally operates at a lower voltage than the utility network. Even for quite large wind turbines (600 kW) it is normal to use generators rated at 690 V. Therefore, as shown in Figures 3.4 and 3.5, a transformer is required to connect each wind turbine to the network or to the wind farm power collection system. With large wind turbines it is usual to provide each turbine with its own transformer to reduce electrical losses.

3.2.4 Braking System

The power in the wind is proportional to the cube of the wind speed and very considerable forces are generated in high winds. It seems obvious that effective braking systems are essential for the safe operation of wind turbines. There are usually at least two independent systems, each capable of bringing the wind turbine to a safe condition in the event of high winds, loss of connection to the network or other emergencies.

With a pitch-regulated rotor, rotating the blades to a zero or negative angle of attack will cause the rotor to slow. This option is not available on a stall-regulated rotor, so some stall-regulated blades have tip brake devices to stop the rotor. Early wind turbines used independent centrifugally acting tip brakes but a common control system is now more usual.

To bring the rotor to a complete stop, mechanical brakes are fitted to the main transmission shafts. It may be desirable to fit a brake to the low speed shaft, so its operation does not depend on the integrity of the gearbox. However, the torques on the low speed shaft can be very large and it is usually much cheaper to use a brake on the high speed shaft. Considerable care is taken to ensure that the braking systems are fail-safe in the event of any malfunction.

3.2.5 Yaw System

A horizontal-axis wind turbine has a yaw system that turns the nacelle according to the actual wind direction, using a rotary actuator engaging on a gear ring at the top of the tower. The wind direction must be perpendicular to the swept rotor area during normal operation of the turbine. A slow closed-loop control system is used to control the yaw drives. A wind vane, usually mounted on the top of the nacelle, senses the relative wind direction and the wind turbine controller then operates the yaw drives.

In some designs the nacelle is yawed to reduce power in high winds, and in extreme conditions the machine can be stopped with the nacelle turned so the rotor axis is at right angles to the wind direction. Although apparently simple, the yaw system has proved one of the more difficult parts of a wind turbine to design. Prediction of yaw loads remains uncertain, especially in turbulent wind conditions.

3.2.6 Tower

The most common types of tower are the lattice or tubular types constructed from steel or concrete. Smaller, cheaper towers may be supported by guy wires. Most modern medium-sized and large wind turbines have tubular towers, which allow access from inside the tower to the nacelle during bad weather conditions. The tower must be designed to withstand wind loads and gravity loads. The nacelle is placed on the top of the tower, and the yaw system allows the nacelle to turn into the wind direction. The tower has to be mounted to a strong foundation in the ground and is designed so that either its resonant frequencies do not coincide with induced frequencies from the rotor or they can be damped out.

A stiff tower is one whose natural frequency lies above the blade passing frequency; the opposite is true for a soft tower. Soft towers are lighter and cheaper but they allow more movement and suffer higher stresses. As the natural frequency of a soft tower is below the blade passing frequency, a transient resonance will be excited each time the turbine is run up to speed. Although this transient resonance generates some movement of the nacelle, it has a short duration and does not cause any difficulty.

3.2.7 Control and Monitoring System

A fully automatic control system is required for the operation and protection of the wind turbine. This must be capable of controlling the automatic start-up, the blade pitch mechanism (on pitch-regulated turbines) and shutdown in normal and abnormal conditions. Besides the controlling function, the system is also used for monitoring purposes in order to provide information on operational status, power production, wind speed and direction, etc. The system will be computer based and, in all except the smallest machines, control and monitoring is possible from a remote location. The control system thus consists of several main functions:

- Sequence control for start-up, shutdown and the monitoring of alarms and trip signals
- Slow closed-loop control of the yaw system
- Fast closed-loop control of the pitch mechanism (if the turbine is pitch regulated)
- Communication with the wind farm controller or remote computer

3.3 OPERATIONAL CHARACTERISTICS

3.3.1 Power performance

The power curve for a wind turbine indicates the net electrical energy output from a wind turbine as a function of the wind velocity at hub height. The curve is determined either by theoretical calculations like those described in Chapter 2 or by field tests. The tests are carried out according to international guidelines and recommendations such as those produced by the International Energy Agency (1990). Power curves from tests are plotted using 10 minute average data of electrical output power and hub-height wind speed measured some distance from the turbine. Averaging of the data masks any transient effects.

The efficiency of a turbine is defined as the ratio of the output of the turbine to the energy of the wind passing through the disc area described by the rotor.

A power curve for a grid-connected, stall-regulated wind turbine is shown in Figure. 3.14; it indicates the following parameters:

- *Cut-in wind speed Vi*: the wind speed at which the turbine starts to produce net power; this is higher than required to start the blades rotating
- *Cut-out wind speed Vo*: to reduce loads on the turbine, It is stopped at wind speeds above the cut-out wind speed.
- *Rated power* P_r: the nominal maximum continuous power output of the wind turbine at the output terminals of the generator (net of losses)
- *Rated wind speed* V_r: the wind speed at which rated power is produced

Figure. 3.14 plots the power in kilowatts and it is the net electrical power available from the generator, allowing for aerodynamic, mechanical and

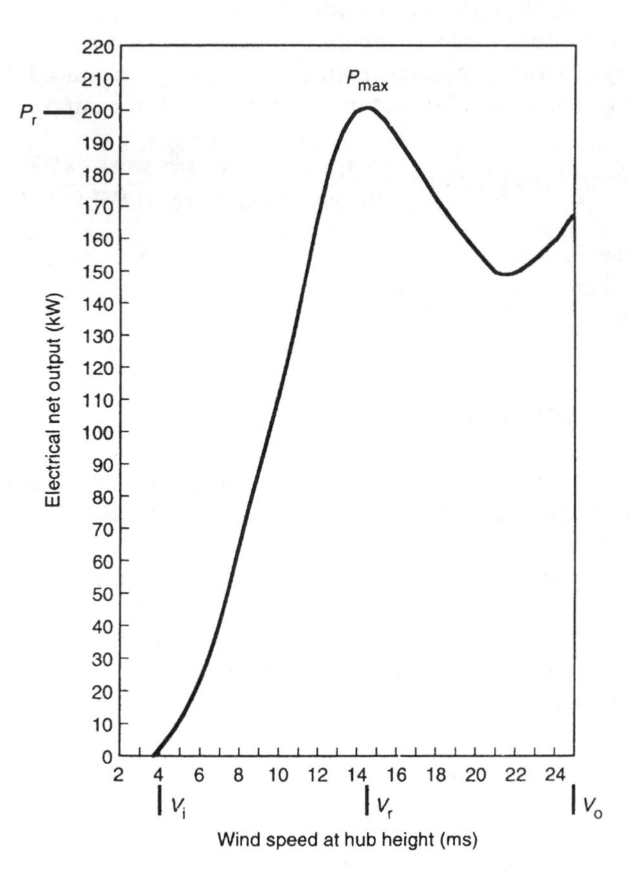

Figure 3.14. Power curve for a 200 kW grid-connected stall-regulated wind turbine

electrical losses. With increasing wind speed the power curve flattens off as the rated power is approached. Ideally the power remains constant at the rated power until the cut-out wind speed, as shown in Figure 1.7. This can generally be achieved with pitch-regulated wind turbines (at least with 10 minute mean data) but Figure 3.14 is the power curve for a stall-regulated wind turbine.

3.3.2 Availability

The availability of a turbine defines the fraction of time when the turbine is available to produce power. Typical availabilities for modern wind turbines are 95–99%, better than many, items of conventional generating plant. The 1%–5% of the time when the turbine is not available is due to maintenance or breakdown.

Useful power is only produced for wind speeds between the cut-in and cut-out wind speeds; depending on the wind conditions, the turbine will be operating at a level lower than its availability.

Another measure is the load or capacity factor, defined as the ratio of the actual energy generated in a time period to the energy produced if the wind turbine had run at its rated power over that period. For example,

$$\text{Weekly load factor} = \frac{\text{energy generated per week (kWh)}}{\text{turbine rated power (kW)} \times 168}$$

There are several similar measures of power plant performance. To avoid confusion when comparing the performance of wind plant, the precise definitions of availability or load factor should be clearly understood.

3.3.3 Annual energy calculation

The calculation of the annual energy yield of a wind turbine is of fundamental importance in the evaluation of any project. The long-term wind speed distribution is combined with the power curve of the turbine to give the energy generated at each wind speed and hence the total energy generated throughout the year. It is usual to perform the calculation using 1 m/s wind speed bins as this gives acceptable accuracy. Using the data of Figure 3.15 the calculation may be tabulated as shown in Table 3.1 and expressed mathematically as

$$\text{Energy} = \sum_{i=1}^{i=\mathrm{n}} \mathrm{H}(i)\mathrm{W}(i)$$

where $\mathrm{H}(i)$ is the number of hours in wind speed bin i and $\mathrm{W}(i)$ is the power output at that wind speed.

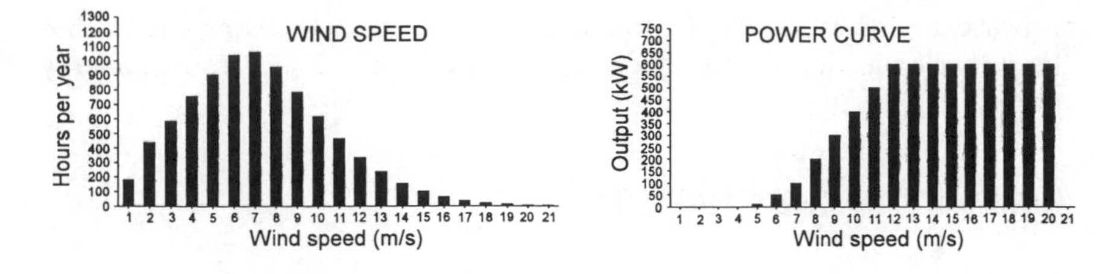

Figure 3.15. Calculation of annual energy production

Table 3.1. Calculation of annual energy yield

Wind speed bin (m/s)	Hours per year	Power output (kW)	Energy yield (kWh)
1	191	0	0
2	444	0	0
3	592	0	0
4	763	0	0
5	913	10	9 130
6	1 037	50	51 850
7	1 058	100	105 800
8	954	200	190 800
9	781	300	234 300
10	613	400	245 200
11	460	500	230 000
12	330	600	198 000
13	235	600	141 000
14	153	600	91 800
15	101	600	60 600
16	61	600	36 600
17	35	600	21 000
18	21	600	12 600
19	12	600	7 200
20	6	600	3 600
Total	8 760		1 639 480

3.3.4 Array efficiency

The output of a wind farm at a given wind speed is less than the output of the same number of isolated wind turbines, this is because the wake effects from upstream wind turbines reduce the power available for those downstream. The turbulence intensity is also increased, a factor which has to be taken into account when considering fatigue life of components and relatively high frequency power variations. Depending on the circumstances, array effects can reduce the output of a wind farm by about 5–15%. In a wind farm there will also be losses within the transformers and, to a lesser extent, within the power collection cables. These will typically be 1–2% of annual energy output.

SUMMARY OF THE CHAPTER

The main components of a horizontal-axis wind turbine for electricity generation were described: the rotor, including the blades, hub and method of power control or regulation; the transmission through the rotor shaft and gearbox to the generator; the nacelle and its mounting and yaw control; the types of tower and the need to take account of the characteristics of the foundations; electrical connections; and the control and monitoring system.

The major parameters associated with the power curve and the operation of wind turbines were described. The power available at the rotor of a wind turbine for a given wind speed is reduced by inefficiencies in the drivetrain, in the generator, in the electrical system and by the array effects resulting from the mutual interference between closely spaced wind turbines in wind farms. The output of a wind farm can be calculated from a knowledge of the wind regime, the aerodynamic characteristics of the rotor, the mechanical and electrical characteristics of the drivetrain, the annual availability of the wind plant and the array efficiency.

BIBLIOGRAPHY AND REFERENCES

Eggleston, D.M. and Stoddard, F.S., *Wind Turbine Engineering Design*, Van Nostrand, New York, 1987.

Freris, L.L. ed., *Wind Energy Conversion Systems*, Prentice Hall, London, 1990.

Hau, E., Langenbrinck, J. and Palz, W., *WEGA Large Wind Turbines*, Springer-Verlag, Berlin, 1993.

International Energy Agency, Expert Group Study on Recommended Practices for Wind Turbine Testing and Evaluation, *Power Performance Testing*, 2nd ed., IEA, 1990

Lipman, N.H., Musgrove, P.J. and Pontin, G.W.W. eds, *Wind Energy for the Eighties*, British Wind Energy Association, Peter Peregrinus, Stevenage, UK, 1982.

McPherson G., *An Introduction to Electrical Machines and Transformers*, Wiley, 1981.

Taylor, R.H., *Alternative Energy Sources for the Centralised Generation of Electricity*, Adam Hilger, Bristol, UK, 1983.

SELF-ASSESSMENT QUESTIONS

PART A. True or False?

1. Most wind farms have turbines rated at over 1 MW.

2. Turbines with single blades have been tested.

3. Rotors should always be mounted on the windward side of the tower.

4. The blades of a pitch-regulated rotor are designed to avoid flow separation.

5. The generator of a horizontal axis turbine is often mounted at ground level.

6. The nacelle of a wind turbine can be rotated independently of the wind direction.

7. The rotor of a wind turbine begins to rotate before the cut-in wind speed.

8. Stall regulation is used to control the power before the rated wind speed is reached.

9. The availability of a wind turbine is given for a specified time period.

10. The output of a wind farm at a given wind speed is the same as adding the output of the individual turbines running on their own at the same wind speed.

PART B

1. Name four of the parameters used to specify a wind turbine.

2. Discuss the factors that determine how many blades a wind turbine should have.

3. Why is pitch regulation not always as reliable as stall regulation?

4. What are the problems in designing the internal transmission system of a wind turbine?

5. Compare the advantages and disadvantages of using synchronous and induction generators in wind turbines.

6. List some of the design considerations relevant to towers for wind turbines.

7. Define the various wind speed parameters associated with operating wind turbines.

8. From the cut-in wind speed of 5 m/s to the speed at which the output first reaches 100 kW,the horizontal-axis turbine in Part B, Question 7 of Chapter 1 has a turbine characteristic, of the form $P = A + BV^3$.
   (a) Calculate the turbine power to the nearest kilowatt at a wind speed of 10 m/s. (Coefficients A and B must be calculated to answer this question.)
   (b) What is the coefficient of performance C_p at a wind speed of 10 m/s? Comment on your answer to this question. What are the implications for the design and operation of this wind turbine?

9. The wind speed at hub height throughout the year varies according to a Weibull distribution of the form $P(V) = \exp\{-(V/C)^k\}$, $P(V)$ is the proportion of the time the wind speed is above V, the shape parameter or characteristic wind speed $C = 7$ m/s; and the Weibull shape parameter, $k = 2$.

(a) For how many hours in the year will the wind turbine in Part B, Question 7 of Chapter 1 operate at full load? (There are 8760 hours in a year.)
(b) How many units (kWh) of electricity will the wind turbine generate at full load?

10. The long-term meteorological statistics give the variation of wind speed in a particular area and can be used for initial feasibility studies. Discuss the significant factors affecting the wind in a particular location. What is the biggest danger in relying on existing statistics?

Answers

Part A

1. False; 2. True; 3. False; 4. True; 5. False; 6. True; 7. True; 8. False; 9. True; 10. False;

Part B

1. Hub height, rotor diameter or swept area, blade solidity, tip speed ratio, rated power, rated wind speed.

2. Having more than two blades reduces cyclic stresses and marginally increases energy capture, but the drag force on the rotor increases and the weight load on the tower increases.

3. The mechanism for rotating the blades is not always mechanically reliable, especially in high wind conditions. However, it should be noted that stall regulation is not very precise.

4. The main problem is designing for the fluctuating torque experienced by the rotor, which can be of a high magnitude. Environmental considerations also require the noise from the transmission system to be kept low.

5. See section 3.2.3.

6. Towers must withstand both wind loads and the weight of nacelle and rotor. To prevent resonant stresses, the natural frequency of vibration for the tower should avoid coinciding with the blade passing frequency over prolonged time periods.

7. Cut-in wind speed is the wind speed at which the turbine starts to produce power. Cut-out wind speed is the wind speed up to which the turbine can operate safely. Rated wind speed is the wind speed at which rated power is produced.

8. (a) $A = -6.033$, $B = 0.04826$, 42.2 kW;

   (b) 0.224

   No calculation is necessary for this question; the answer can be deduced. Since the power increases as the cube of the wind speed, C_p is almost constant (neglecting the small offset due to the cut-in wind speed), i.e. the machine always operates near maximum efficiency. It must therefore operate at variable speed. The designer of this machine could have taken the opportunity to raise C_p, perhaps by improving the blade profile.

9. (a) 3.2% or 278 h; (b) 27 800 kWh

10. Review the whole of Chapter 1 before answering this question. The biggest danger in relying on existing statistics is to overestimate the performance of the wind turbines; this would have an adverse effect on the economics of the project.

PART **B**

Project Assessment and Engineering

Understanding the wind and its conversion into useful energy allows the performance of a particular type of wind turbine to be calculated and hence its ability to provide useful energy. However, this is only part of the story. How much will the wind turbine cost? What will be the value of the energy produced and how will it be used? How will the project be financed? Are environmental questions important?

Planning a wind project, using wind turbines of known cost and performance, requires consideration of economics and environmental issues. A wind farm will require connection to the electricity network. Is the local distribution network suitable for the connection? What will be the effect of the wind farm on the power quality and the electrical losses of the network?

The design of a windpump installation needs a knowledge of the water-pumping requirements. Again, what measures are needed if the wind turbine is not operating? Can pumping be carried out by hand or would a stand-by pumping engine be needed?

Effective management is essential for the successful realisation of wind projects; modern management tools may be required for complex projects.

This part of the book is concerned with the way wind turbine projects are assessed, planned and managed. The term *planning* covers all aspects of the project programme, including those concerned with obtaining the necessary authorisations, sometimes known as planning permission. The chapters should not be viewed in isolation, each one depends to a greater or lesser extent on the others. For example, the importance of environmental impact is discussed in Chapter 5, but environmental factors also have a bearing on the economic assessments in Chapter 4.

4

Economic Assessment

AIMS

This unit outlines the principles underlying the economic assessment of wind power plant. It explains the need to take account of all costs related to wind turbine construction and operation in an economic model, including initial capital cost, annual running costs (which include maintenance and operation) and capital replacements such as major spares. The main method of economic assessment, discounted cash flow, is described in detail. Various methods of measuring the value of a project are described, including simple payback, net present value, internal rate of return and benefit-to-cost ratio. Although the unit concentrates on wind turbines for electricity generation, the principles of economic assessment can be applied to the use of wind plant for any purpose.

OBJECTIVES

After completing this unit, you will have an understanding of the following five areas:

1. The ideas underlying economic appraisal.
2. Economic objectives.
3. Economic methodology.
4. Making economic comparisons of different wind power options.
5. The need to take non-economic factors into account in project appraisal.

NOTATION AND UNITS

n	Number of years
r	Interest or discount rate
A	Annualised value or constant yearly payments
B/C	Benefit-to-cost ratio

IRR	Internal rate of return
NPV	Net present value, i.e. the difference between cost present value and benefit present value
PV	Present value

4.1 ECONOMIC CONSIDERATIONS

The factors which have to be taken into account when appraising the economics of power generation are complex and interrelated. However, it is possible to gain an insight into the principles of economic appraisal by consideration of some simple examples of alternative methods of electricity generation. The examples considered below are hypothetical but realistic: a nuclear station, a coal-fired station, and wind turbine plant. This unit seeks to demonstrate that the lifetime costs of each alternative must be considered when a supplier of electricity, or an electricity utility is carrying out an economic appraisal and this will include some less obvious costs concerned with the rest of the electricity generating system. Non-economic constraints must also be considered, which may be environmental, ecological, social or political.

In pursuit of its economic objectives, an electricity utility must secure the best use of its resources by selection of the project that gives the best value for money, within constraints. An economic model needs to be developed, using a set of investment appraisal techniques, which contain the following key activities:

1. Specify the objectives of the investment.
2. Identify possible alternatives, which should include 'do nothing'.
3. Identify the costs and benefits of the alternatives which will include the following:

- Initial capital cost, or the cost of buying the plant
- Annual fixed costs such as numbers of staff or insurance
- Annual variable costs such as fuel or lubricating oil
- The benefits or savings which would be produced by the alternative

4. Identify non-economic factors.
5. Consider the uncertainties involved in the project.

Note that under item 3, the initial capital cost is the cost of buying the plant. Annual fixed costs do not depend on how much the plant is run, e.g. the cost of employing permanent staff or carrying public liability insurance. Annual variable costs depend on plant operation and the costs of fuel if appropriate, lubricating oil or consumable spares. There is no hard and fast division between any of these classes. The present value (PV) of an annual cost can be regarded as a capital cost (see below).

4.2 METHODOLOGY

This discussion assumes there are several options under consideration for a particular project. Is one wind turbine required, or several? Is wind power the only way of fulfilling the requirement, or are there others such as, diesel-powered plant? The benefits and costs of each alternative need to be considered. This unit describes how to determine which project would be best, taking into account economic and non-economic factors, and should therefore be chosen.

What is an economic model? It is a mathematical function which has financial input and output variables. It may consist of a simple model of a particular type of generator, or it may seek to model the entire system, in which case it will be the very opposite of simple. To illustrate what is involved we will work through a number of simple comparisons, and look at the results of a much more complex modelling activity.

4.3 DISCOUNTING

An immediate problem is how to deal with alternatives having completely different combinations of capital and annual costs. The solution is to 'discount' all costs in the future to a fixed starting date to derive a single present value (PV) of the project. This is very simple in principle, but initially it may appear complicated. It works on the assumption that money in your pocket (i.e, what you have now) is worth more than the same amount of money you may expect to receive in the future, say, one year hence. This future money has to be discounted to produce its PV using an interest rate, or discount rate. For example, the present value of \$1 which you might earn a year from now, if the interest rate is 10%, is \$1/1.1, and the present value of \$1 two years hence is $\$1/1.1^2$ and so on. This applies to payments as well as income.

Now consider the case where you pay \$$A$ at yearly intervals for a period of n years. What is the present value of this benefit stream (cost to you, benefit to someone else)?

$$\text{PV} = A\left(\frac{1}{1+r} + \frac{1}{(1+r)^2} + \frac{1}{(1+r)^3} + \cdots \frac{1}{(1+r)^n}\right) \tag{4.1}$$

The right-hand side of equation (4.1) is a geometric series that sums to

$$\text{PV} = A\left[\frac{(1+r)^n - 1}{r(1+r)^n}\right] = \frac{A}{r}[1 - (1+r)^{-n}] \tag{4.2}$$

This is a very important expression. To illustrate its significance, consider the case where you *borrow* \$30 000 from the financing institution at 8% interest over 25 years. The PV is the amount you borrow, $n = 25$ years, $r = 8/100$, A is your annual repayment.
Rearranging equation (4.2) and substituting these values gives

$$A = \text{PV} \times \frac{r(1+r)^n}{(1+r)^n - 1} = 30\,000 \times \frac{0.08 \times 1.08^n}{1.08^n - 1} = \$2810 \text{ per year} \qquad (4.3)$$

So you can see how \$30 000 is equivalent to \$2810 per year for 25 years. Discounting \$2810 for 25 years at 8% gives a PV of \$30 000. Try this if you wish by substituting the values in the right-hand side of equation (4.2).

This is the method of investment appraisal used by governments and largc organisations and recommended by the International Energy Agency (1991) for the evaluation of renewable energy projects, particularly those which are complex and costly. A valuable treatment of the economics of windpumps is given by Fraenkel *et al.* (1993).

Discounting can be used either to derive the PV or the annualised value (A), which will take into account *all* expenditures that occur during the lifetime of the project, whether they are initial capital cost, regular annual costs or occasional costs (e.g. replacement of a major capital item after 20 years). We can consider the PV of a project as the amount of money we would pay to buy the equipment and to pay somebody a lump sum to operate and maintain it. We can look at it another way: A is the annual amount we would pay for the project without an initial capital cost, i.e. the annual amount required to *lease* the project.

4.4 MEASURING THE VALUE OF AN INVESTMENT

The benefits or savings from the project also need to be taken into account, so their PV is also calculated. The difference between the PVs of costs and benefits is the *net present value* or NPV. NPV is not the only measure of economic worth. There are others, including

- internal rate of return (IRR)
- benefit-to-cost ratio (B/C ratio)
- simple payback period

The IRR is the value of discount rate at which the project makes neither profit not loss. The B/C ratio is the ratio of the discounted benefits to discounted costs. If the ratio is greater than unity, the project clearly makes a profit.

The economic worth of a project may be indicated approximately by a simple payback period, which is merely the time from the start of the project to when it breaks even, i.e. when costs exactly balance benefits. After this time, the project is making a profit. The initial investment is offset against the net monthly or annual savings. For example, a project may produce a net saving of \$100 per month against an initial investment of \$1500. The simple payback period is 15 months. This method is useful as an approximate measure for projects that are similar and have short payback periods, but it can give misleading results, particularly when there are wide differences in the cost and benefit streams. It should therefore be used with caution.

4.5 ALLOWING FOR INFLATION

The advantage of discounting is that it removes the need to take account of general inflation. However, if a rise in the value of a commodity relative to everything else is expected – *real* rise in its cost – then an *inflator* must be used. An example is the price of fossil fuels, which will rise when the supply becomes less than the demand.

In addition, it is extremely important to relate each price to a common date. For example, if all costs of a project are to be discounted to mid-1995 then the cost of an item relating to mid-1990 must be adjusted to take account of inflation over this period. Price indexes can be used for this purpose. For example, suppose the price index for mid-1993 is 100 and mid-1995 prices are 12.5% higher. The price index for mid-1995 is therefore 112.5. All mid-1993 prices must be increased by 12.5% if the discounting date is mid-1995. Conversely, mid-1995 prices must be *reduced* by 12.5% if the discounting date is mid-1993. This is a particularly important consideration in times of high inflation.

4.6 EXAMPLES

The preceding sections have concentrated on ideas underlying economic appraisal. This section develops some of those ideas by discussing examples concerned with nuclear, coal and wind powered electricity generation. Break-even costs and their implications are considered. Table 4.1 shows the basic data for these examples. The capital cost of a piece of equipment is the amount that you would have to pay to have it installed and ready to work on a given date. It is conventional to present the costs as if they were actually paid on the very day the plant started up. In practice this does not usually occur, particularly in large capital projects when capital costs are paid in a number of stages. Some of these payments may be made before the plant is commissioned, and some after. These payments are discounted to the date of commissioning; the sum of their PVs is the single capital cost we are considering. This sum will be greater than the undiscounted sum; the ratio of the sum of the PVs to the undiscounted sum is called interest during construction (IDC).

Table 4.1. Example cost and performance figures for electricity generating plant

Costs	$	Nuclear	Coal	Wind
Capital	(\$/kW)	1500	1000	600–1500
Fixed	(\$/kW pa)	22	15	4
Fuel	(¢/kWh)	0.5	1.7	0.1
Variable	(¢/kWh)	0.2	0.2	0.2
Load factor	(%)	75	75	25
Life	(years)	40	40	25

For example, suppose the supplier of the coal-fired station quotes a price of \$818/kW not including interest during construction. It is agreed that payments will be made as in columns 1 and 2 of Table 4.2. The PV of each payment is calculated as in section 4.4. For example, the PV of 25% of \$818/kW paid 8 years before commissioning, with a 5% discount rate, is

$$818 \times 25/100 \times 1.05^8 \$/\text{kW}$$

Table 4.2. Calculation of interest during construction

Percentage of capital cost	Years before (−) or after (+) commissioning	PV of cost at commissioning (\$/kW)
25	−8	302
50	−4	497
15	0	123
10	+1	78
	PV of capital cost	1 000

The value of interest during construction (IDC) is calculated as follows:

$$\text{IDC} = (1000 - 818) \times 100/818 = 22.2\%$$

A further factor which has to be considered is whether capital will have to be spent during the life of the project after it has been operating for a number of years. As an example, consider the wind power capacity, where we have assumed a life of 25 years, in contrast to the 40 years assumed for coal and nuclear plant. We must either buy a completely new replacement or we must refurbish the old wind plant; each will require capital expenditure at year 25. This expenditure is discounted to the start date of the project.

This introduces a further complication. Assuming the refurbished wind power plant will have a life of 25 years, it will have a remaining life of 10 years when the nuclear and coal plants come to the end of their 40 year lives. We then have to ascribe some capital value to wind turbines. This introduces the idea of depreciation. Assuming the value depreciates linearly with time, we can work out the remaining capital value of the wind turbine plant, which is discounted and taken as a credit when working out the total PV of the project. This is worked out as in Table 4.3.

Table 4.3. Dealing with capital costs incurred during the life of the project

A	Initial capital cost	\$600/kW
B	Cost of refurbishment (50%)	\$300/kW
C	PV of this cost	$300/1.05^{25} = \$89/\text{kW}$
D	Capital value at year 40	$300(1-15/25) = \$120/\text{kW}$
E	PV of D	$120/1.05^{40} = \$17/\text{kW}$
	Net capital PV (A + C − E)	\$672/kW

A final factor to be taken into account is that 1 kW of capacity of one type of generation will not in general produce the same number of kilowatt-hours per year as other types of plant. For simplicity, we have assumed that nuclear and coal produce the same number of units per year, so they have the same annual load or capacity factor as defined in section 3.2.2. This will not be the case for wind power plant, which will have an annual load factor of about 25%. The wind power capacity required to produce the same number of units as the other plant must have, on our assumptions, three times the capacity of the coal or nuclear plant. This is because it produces one-third of the electricity in a year, as indicated by the load factors.

The capital cost of wind power plant to produce the same energy as nuclear or coal plant is therefore

$$672 \times 75/25 = \$2016/\text{kW}$$

The fixed annual costs are the total of all 'fixed' costs per year. This means simply all costs which do not change, however much the plant operates. They include rent, rates, insurance, research and development, salaries and training. An everyday example is the insurance on a car. The PVs of the fixed annual costs are calculated using equation (4.2), where A is the total of fixed annual costs per year. For a discount rate of 5% ($r = 0.05$) and $n = 40$ years, PV $= 17.159A$. So if $A = \$4/\text{kW}$ for wind power, PV $= 17.159 \times 4 = \$69/\text{kW}$ of wind power capacity, or $3 \times 69 = \$206/\text{kW}$ of capacity equivalent to coal or nuclear plant.

The calculations for the nuclear and coal plant are similar in principle and are given in Table 4.4.

The variable annual costs are those which change according to the amount the plant operates. They will include costs of lubricating oil, replacement parts and inspection for all types of plant, fuel and ash handling in the case of coal plant, nuclear fuel reprocessing costs for nuclear plant. Tyres for a car is an everyday instance of a variable cost.

Table 4.4. PV $ of costs for electricity-generating plant

Cost($/kW)	Nuclear	Coal	Wind
Capital	1 500	1 000	2 016–5 034
Fixed	377	257	69(×3)
Fuel	565	1 922	–
Variable	225	225	75
Total	2 667	3 404	2 297–5 315

As an example, the fuel costs for coal-fired plant are 1.7c/kWh of electricity generated. The total annual fuel cost is calculated as follows:

Annual fuel cost $= 1.7 \times 8760 \times 75/(100 \times 100) = \$112/\text{kW}$
PV of fuel cost $= 112 \times 17.159 = \$1922/\text{kW}$

The PVs of each type of expenditure are given for each project in Table 4.4. The sum of the PVs of the projects are the figures we compare to determine which is the most economic. We can see that nuclear plant is the cheapest if the cost of wind power is at the top end of its range. This is not so if the cost of wind power is at the bottom of its range, then it becomes the cheapest option.

From Table 4.4, wind power plant must have a total PV of $2667/kW to be competitive with nuclear plant, or $3404/kW to be preferred to coal. The equivalent capital costs for wind, taking into account its lower annual load factor, work out to be $841/kW and $1087/kW respectively.

Finally, we can express the total cost of each project for comparison in other ways. One way is the total cost in $/kW per year, when the initial capital cost is annualised and added to the other annual costs. Another, and obviously attractive way, is to calculate the cost per kilowatt-hour, by dividing the total annualised cost by the units generated per year, which is the product of the number of hours in the year and the load factor per kilowatt of installed capacity. For example, for nuclear plant:

Annualised total PVs $= 2667/17.159 = \$155/\text{kW}$ per year
Total cost per unit $= 155 \times 100/(8760 \times 75/100) = 2.4$ ¢/kWh.

The costs for all the technologies are given in Table 4.5.

Table 4.5. Comparison of PVs, annualised costs and unit cost

Cost	Nuclear	Coal	Wind
Annual cost ($/kW per year)	155	198	134–310
Unit cost (¢/kWh)	2.4	3.0	2.0–4.7
PV ($/kW)	2667	3404	2297–5315

4.7 INFLATION AND LEVELLISED PRICES

So far, the calculations have assumed that inflation is uniform for all prices considered, so the economic rankings are not affected, whatever the rate of inflation. However, we all know that fuel prices rose sharply in real terms after 1974. A fourfold increase in oil prices almost overnight produced increases to a lesser extent in the price of coal, because of the rise in demand for it. It seems obvious that if the real price of a commodity rises significantly by comparison with another, the economics of a project will be affected if it needs the commodity. To take this into account, we need to introduce an inflation rate into the discounting procedures. This can be done simply by adjusting the annual cost streams appropriately. Assuming a constant real rise in coal prices (2% per year) and nuclear prices (1% per year) relative to those of all other costs, the figures in Table 4.5 become as shown in Table 4.6.

Not surprisingly, the competitive position of wind is improved against both nuclear and coal plant, and the economics of coal plant are worsened by comparison with nuclear because a more rapid rise was assumed (for the sake of argument) in the real cost of coal.

Table 4.6. Comparison of PVs, annualised costs and unit costs including inflation in fuel prices

Cost	Nuclear	Coal	Wind
Annual cost ($/kW per year)	160	235	134–310
Unit cost (¢kWh)	2.4	3.6	2.0–4.7
PV ($/kW)	2752	4038	2297–5315

These results show that lower cost energy sources improve their economic position in times of high fuel cost inflation, as is intuitively obvious. A technology may have high capital costs, like wind power, but it can be a very good investment in times of high fuel cost inflation. Consideration also needs to be given to situations where real prices may fall. Fossil fuel prices in some countries are relatively low at present, so renewable energy schemes of high capital cost are less competitive.

4.8 NON-ECONOMIC FACTORS AND EXTERNAL COSTS

Energy developments on a large scale, which include wind power projects, can have a profound effect on society and this may not always be beneficial. Environmental, social, political, and aesthetic factors must be considered. Some may prevent a project being realised, such as prohibition of certain types of development in a national park.

Other factors may increase or reduce the cost of projects. The economic objectives of the project must be sought within a framework imposed by these non-economic factors.

This brings us to the concept of external costs. A power station may produce emissions which are damaging. Acid rain may cause damage to buildings which can be quantified in terms of repair or maintenance costs.

Carbon dioxide will contribute to the greenhouse effect. The effects of damaging emissions on human health may be reflected in the costs of medical treatment. Other types of generator, such as wind turbines, will not produce emissions, but may cause loss of visual amenity that is difficult to quantify. Costs that are not incorporated into the cost streams of a project are called *external* costs.

There is an increasingly strong body of opinion that the external costs of a project should be *internalised* and should therefore be included in the cost streams of the project. A carbon tax on fossil fuel (which increases annual variable costs) is an example of how governments can internalise the effects of carbon dioxide emissions. The project developer could counter the effects of the carbon tax by taking steps to increase the plant efficiency to reduce fuel usage and reduce liability for carbon tax. However, this would require an increase in capital cost. The overall effect is to improve the economic position of non-polluting energy sources such as wind or hydropower, thus encouraging their use.

SUMMARY OF THE CHAPTER

This chapter has outlined the different factors which must be taken into account when carrying out economic assessments of power generation projects. It draws attention to the need to consider initial and other capital costs, and annual costs which include maintenance and operation. Although the simple payback period may be a useful measure of the short-term economic performance of similar projects, the need to use discounted cash flow techniques is emphasised for long-term capital projects.

BIBLIOGRAPHY AND REFERENCES

International Energy Agency, *Guidelines for the Economic Analysis of Renewable Energy Technology Applications* IEA, Quebec, 1991.

Fraenkel, P. Barlow, R., Crick, F., Derrick, A., and Bokalders, V. *et al*, *Windpumps: A Guide for Development Workers*, Intermediate Technology Publications, London, 1993, 71.

SELF-ASSESSMENT QUESTIONS

PART A. True or False?

1. The wind is free so wind power costs nothing.

2. A dollar in your pocket is worth exactly the same to you as a dollar you receive 20 years from now.
3. At 10% interest rate, a dollar given to you a year from now is worth about 91% of the dollar you have in your pocket.
4. Fixed annual costs include the cost of employment of permanent staff and insurance cover.
5. Variable annual costs do not change in accordance with the operating regime of the plant.
6. The internal rate of return (IRR) is the value of the discount rate at which a project makes neither profit nor loss.
7. The internal rate of return (IRR) is the value of the discount rate at which the benefit-to-cost (B/C) ratio is more than unity.
8. External costs are incurred when providing weather protection for wind turbines.
9. Wind turbines and other types of renewable energy plant have better economic prospects when external costs are taken into account.
10. Times when fuel cost inflation is high favour the economics of wind turbines.

PART B

1. What is the purpose of an economic appraisal when carried out by an electricity utility or supplier?
2. What is the present value (PV) of $1 two years in the future if the interest rate is 15%?
3. What is the PV of $1 two years in the past if the interest rate is 15%?
4. What is the PV of $2000 for 25 years at 5%?
5. Explain the terms NPV and B/C.
6. Derive a relationship between NPV and B/C.
7. What effect would a real fall in oil prices have on the prospects for wind power?
8. Give some examples of external costs.
9. How could taking account of external costs benefit wind power?
10. Why does wind power assist in the mitigation of the greenhouse effect?

Answers

Part A

1. False; 2. False; 3. True; 4. True; 5. False; 6. True; 7. False; 8. False; 9. True; 10. True.

Part B

1. To secure the best use of the utility's resources by selection of the alternative that gives the best value for money.

2. The PV of $1 two years from now if the interest rate is 15% is $1(1.15)^{-2}$ or 0.756.

3. The PV of $1 two years in the past if the interest rate is 15% is $1(1.15)^{2}$ or 1.3225.

4. Using equation (4.2), PV = $2000 \times (1.05^{25} - 1) / 0.05\ (1.05)^{25}$ = $28 188.

5. NPV, or net present value, is the difference between the PV of the costs and benefits of a project. B/C, or benefit-to-cost ratio, is the ratio of discounted total benefits and discounted total costs of a project.

6. If PVC is the present value of cost and PVB is the present value of benefits, B/C = PVB/PVC = (NPV + PVC)/PVC.

7. A real fall in oil prices will reduce the savings to be made by wind power and thus its competitiveness.

8. Examples of external costs are damage to buildings caused by acid rain, and health problems caused by particulate or gaseous emissions from power processes.

9. Wind power does not produce gaseous or particulate emissions, so it does not cause any of the damage in the answer to question 8.

10. The lack of carbon dioxide emissions from wind power generation more than outweighs any emissions caused during the manufacture or erection of wind turbines. Wind turbines therefore help to reducing the rate of increase of carbon dioxide emissions into the atmosphere.

5

Planning Authorisation, Environmental and Social Issues

AIMS

This unit sets out some of the planning authorisation, environmental and social issues relating to wind power installations. These matters need to be considered carefully when appraising both small- and large-scale wind projects.

OBJECTIVES

When you have completed this unit you will have an appreciation of three things:

1. The key issues relating to planning authorisation for wind power installations.
2. The environmental considerations relating to wind power.
3. Social considerations influencing development of wind power.

5.1 PLANNING AUTHORISATION FOR WIND POWER

Planning issues relating to wind power assume their greatest importance in countries where developments are controlled by more or less rigorous planning laws. This is most likely to be the case in developed countries, particularly those of high population density. Important factors are as follows:

1. The existence of a national planning system which constrains industrial and rural development.
2. Guidelines on how developers should obtain planning permission for wind power developments.

3. The attitude of the general public to renewable energy development in general and to wind energy in particular.
4. Development of planning policy relating to wind energy, particularly factors relating to the environmental benefits of renewable energy.
5. The effect of planning requirements on the cost of wind energy.
6. The costs and savings relating to the integration of wind power into existing energy supply networks.

In regions of low population density formal planning issues may be of less consequence. However, careful consideration still needs to be given to the environmental impact and infrastructure required for all wind developments.

An example of advice on planning issues in a developed country is provided by the British Wind Energy Association (1994) in *Best Practice Guidelines for Wind Energy Development*. This refers to the way formal government policy is set out in UK government circulars and planning policy guidelines, such as Planning Policy Guideline Note PPG 22 (Department of Trade and Industry, 1993).

5.2 ENVIRONMENTAL IMPACT

The environmental benefits of wind power result mainly from reduction in the use of fossil fuels, leading to a reduction in emissions of pollutants created by combustion.

- Emission of gases
- Fly ash
- Slag

Perhaps the most important of these is the absence of gaseous emissions, such as oxides of carbon, sulphur and nitrogen. Wind plant does not emit carbon dioxide, so it does not contribute to climatic change (sometimes known as the greenhouse effect). Many conventional fossil fuel generating stations emit sulphur and nitrogen oxides; these contribute to the acid rain effect, which has caused significant environmental damage in Europe. The precise savings in gaseous emissions will depend on the mix of generating plant in the power

Table 5.1 Savings in gaseous emissions from wind generation under typical UK conditions (Reproduced by permission of the British Wind Energy Association)

Emission	Oxide saving per kWh of electricity (g)	Annual oxide saving for 5 MW wind farm (tonne)
Carbon dioxide	800	10 500–16 100
Sulphur dioxide	10	150–240
Nitrogen oxides	3.4	50–80

system to which the wind farm is connected. For the average UK plant the annual savings in gaseous emissions are shown in Table 5.1.

Any emissions caused during the manufacture of wind turbine plant are offset after a few months of emission-free operation. Similarly the energy expended in manufacturing a wind turbine is paid back after about a year's operation (World Energy Council, 1994).

Environmental disbenefits include visual intrusion, land use, impact on local ecology, noise and effects on radio communications and television reception. However, in most cases they can be alleviated by careful siting and design of wind plant, and by a policy designed to secure the maximum public acceptance of wind energy.

Visual intrusion on the landscape is usually the most frequent public objection to wind turbines. The windiest sites are often the most beautiful and have the least sign of human interference. The public's response to visual impact is usually conditioned by attitudes to different types of energy production and the advantages or disadvantages people see for themselves and their society. The best way to present the benefits of wind power is to demonstrate properly sited, well-maintained wind turbines that operate efficiently. The size, height, colour, material, wind farm configuration, landscape background and siting to reduce lines of sight are all design factors that should be taken into account to reduce visual impact.

Wind farms have the advantage of *dual land use*. Wind energy is diffuse and the large-scale capture of energy from the wind requires turbines to be spread over a wide area. A modern wind farm uses 1% of the total land area it occupies and the wind turbines and their bases occupy only 0.2% (Garrad, 1991). The remaining land can be used for agriculture and wind farm access roads are often seen by farmers to be an asset rather than a disadvantage, provided they are properly planned.

The extent of possible disturbance to the local *ecology* caused by both construction and operation of a wind energy project is a concern of many people. The period of construction for a wind farm is very short by conventional power station standards and is typically less than six months. The process disrupts a small proportion of the land area designated for the plant; afterwards the site can be returned to its former condition except for the small area required for the turbines and access roads.

The impact of wind turbines on *bird life* is an important matter and has been surveyed across several countries by Benner (1993). Birds can be affected in two ways: by injury or death following collision with the blades or tower, or by being disturbed in their breeding, nesting or feeding habits. Bird victims can number from zero to several hundred per turbine per year in the most extreme situations. Generally, the numbers of bird victims are less than those caused by equivalent lengths of high voltage lines. One study concluded that for the 3 500 wind turbines in Denmark some 20 000–25 000 birds die yearly from collision with wind turbines, whereas over a million birds are killed by traffic (Clausager,1996) The disturbance to breeding birds appears negligible but is more serious to resting and migratory birds. The siting of wind turbines should be avoided on migration routes and where there are high densities of nesting or foraging birds. Careful detailed design of the turbines and towers can reduce the perching and nesting opportunities for birds.

Like visual impact, objections to *noise* from wind turbines has a subjective element. What is acceptable in an industrial area is not acceptable in a rural area, and what is acceptable during the day is not acceptable at night. Paradoxically the noise from a wind turbine may be more obvious at low wind speeds than at high speeds because the general background noise caused by the wind, such as from trees, is proportionally less. Noise from wind turbines arises from aerodynamic noise from the rotor and mechanical noise from the gearbox, generator, etc. Gearbox noise, usually emitted as discrete tones, can exceed aerodynamic noise but by careful design mechanical noise can be reduced to acceptably low levels. Aerodynamic noise is usually broadband and arises from the fluctuating and unsteady wind profile, and the shedding of vortices from the trailing edge and the tips of the blades.

Noise can be reduced by keeping trailing edges thin and profiling the blade tips. Recent research in this area is described by Hagg *et al.* (1993). A typical modern wind turbine of 300 kW operating at a wind speed of 8 m/s produces a noise level of 45 dBA at about 200 metres (Figure 5.1). Danish regulations

WIND TURBINE NOISE PATTERN

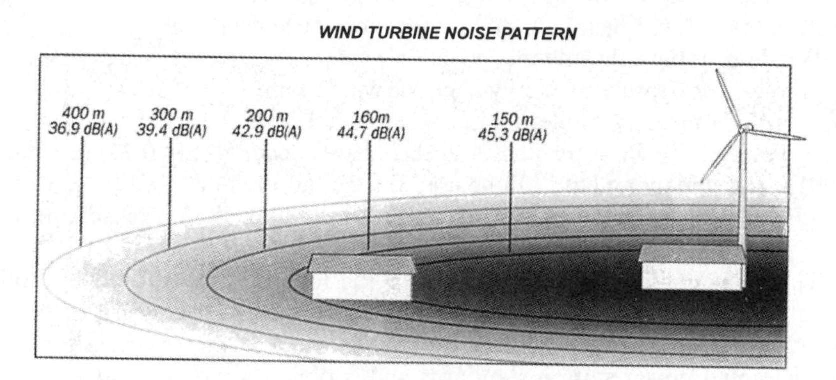

RELATIVE NOISE LEVELS

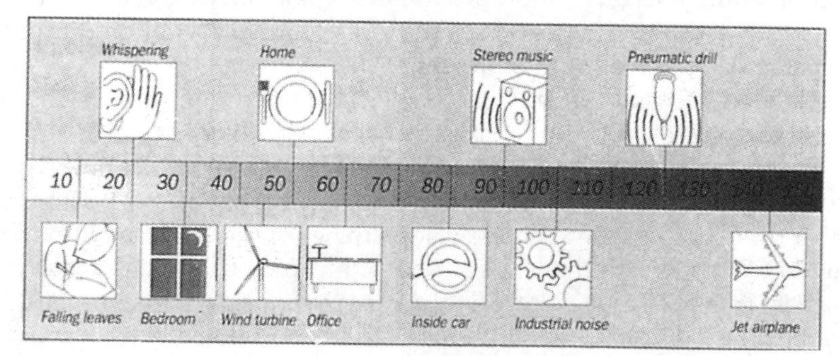

Figure 5.1. Pattern of wind turbine noise (sound pressure level): results of actual measurement (Reproduced from A.D. Garrad, 1991, *Time for Action: Wind Energy in Europe*, by permission of the European Wind Energy Association)

suggest this should be the minimum distance that a wind turbine is situated from houses in order that the noise level is acceptable, and for wind farms the distance should be increased to 500 m (Garrad, 1991). An alternative limit of acceptability is that the noise level with the wind turbines in operation should be maintained within 5 dBA of the existing evening or night-time background noise level.

5.2.1 Quantitative Description of Wind Turbine Noise

Quantitative description of wind turbine noise is an optional part of the course

Unlike visual impact, noise can be quantified. Two distinctly different measures are used to describe wind turbine noise. These are the *sound power level*, L_w, of the source and the *sound pressure level*, L_p, at a location. Both L_w and L_p are normally expressed in dB (decibels). The *sound power level* describes the power of the source of the noise whereas the sound *pressure level* describes the noise at a remote point.

The *sound power level*, L_w, of a source (i.e. the wind turbine) is expressed as

$$L_W = 10 \log_{10}\left(\frac{P}{P_0}\right) \tag{5.1}$$

where P is the sound power level of the source and P_0 is a reference sound power level, often taken to be 1 pW or 10^{-12}W. A typical value of the sound pressure level declared by manufacturers of modern large wind turbines might be in the range 95–105 dBA. The sound power level will vary with wind speed and therefore with the operating conditions of the turbine. The noise created by high wind speeds passing over bushes and trees tends to mask the noise from wind turbines. The most critical wind speeds for noise assessment therefore tend to be around cut-in and it is common to declare the sound power level at a wind speed of 8 m/s.

The *sound pressure level*, L_p, of a noise is defined as

$$L_p = 20 \log_{10}\left(\frac{P}{P_0}\right) \tag{5.2}$$

where p is the root mean square (RMS) sound pressure level and p_0 is a reference sound pressure level often taken to be 20μPa or 20×10^{-6}Pa. A typical value of sound pressure level 350 m from a wind farm would be 35–45 dBA.

It is common to weight the measurements to reflect the sensitivity of the human ear and this is done by applying a filter of the form shown in Figure 5.2. Measurements made with this filter are referred to as dBA or dB(A).

Having obtained the sound power level from the manufacturer, the user of the wind turbine will wish to determine the sound pressure levels around the wind farm to ensure that planning requirements are met and that no nuisance is caused to local inhabitants. One simple model for the propagation

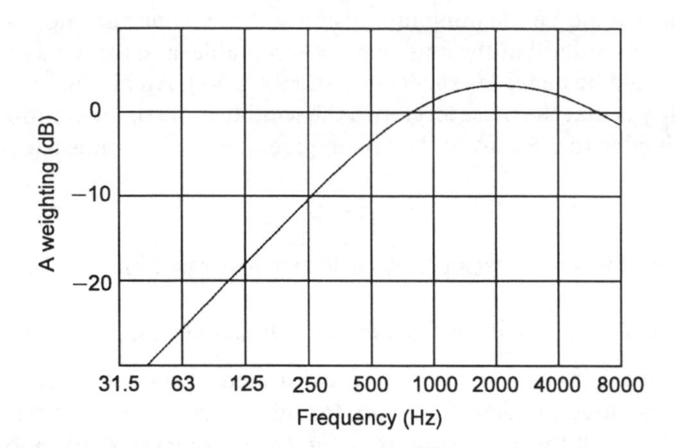

Figure 5.2. dBA weighting function

of noise from a single wind turbine is given by the International Energy Agency (1994) as

$$L_p = L_w - 10\log_{10}(2\pi R^2) - \alpha R \tag{5.3}$$

where L_p is the sound pressure level a distance R from a noise source of sound power level L_w; α is a frequency-dependent sound absorption coefficient; and both L_p and L_w are expressed in dBA. If the noise has a particular tone, e.g. from the gearbox, it is more easily detected by the human ear, so an additional 5 dBA is often added to the sound pressure level for the purpose of assessing acceptability.

Because the decibel scale is logarithmic the addition of two sound pressure levels is carried out like this:

$$L_{1+2} = 10\log_{10}(10^{\frac{L_1}{10}} + 10^{\frac{L_2}{10}}) \tag{5.4}$$

This means the addition of two equal values does not produce a doubling of the sound pressure level but an increase of 3 dBA.

Background noise is a further complication, so it may be necessary to conduct a survey of the ambient noise at adjacent dwellings, without the wind farm in operation. As the background noise is not constant this sound pressure level is often declared as L_{90} the noise level exceeded for 90% of the time.

The subject of noise from wind turbines is extremely complex and regulations for acceptable limits differ from country to country. This brief section has merely introduced some of the basic ideas; Wagner et al. (1996) provides a comprehensive treatment of the subject.

5.2.2 Electromagnetic Interference and Turbine Safety

Disturbance to *telecommunications* like radio and TV by wind turbines arises from electromagnetic interference and appears to be strongly site dependent.

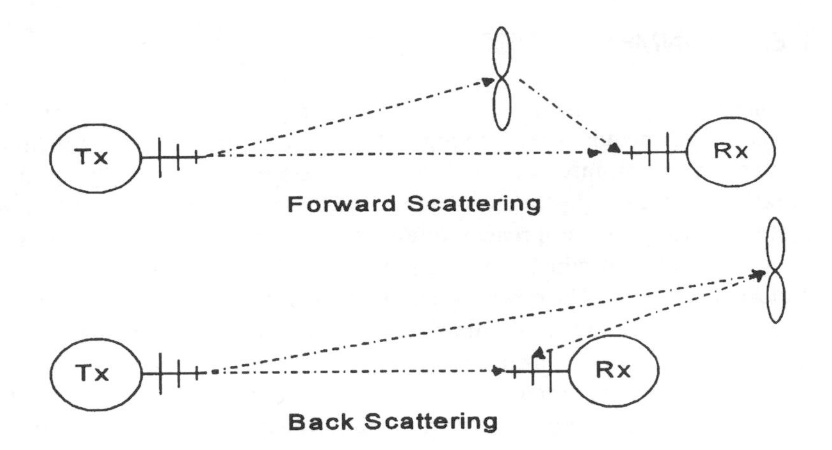

Figure 5.3. Interference mechanisms

The interference occurs because the plane of the rotor disc acts like a mirror; it reflects signals from the receiver and these reflected signals can affect the direct signal to the receiver.

There are two main mechanisms for interference: forward scattering and back scattering (Figure 5.3). Forward scattering occurs when the wind turbine is located between the transmitter and receiver and, for TV signals, it causes fading of the picture at the rotational speed of the blades. Back scattering occurs when the turbine is behind the receiver and gives rise to double or ghost images on the TV screen. The worst conditions exist at high frequencies and when weak signals interact with a large turbine with metallic blades on the line of sight between a poor quality receiver and the transmitter. Careful siting of turbines to avoid microwave routes, and installation of local amplifiers or cable connections to TV sets will usually overcome the interference problem in affected houses, generally few in number.

The *safety* record of wind energy technology is very good; there are no reported incidents concerning members of the public. There have been few serious accidents to staff working on wind turbines and those causing human injury or death have generally been the result of poor management or non-observance of safety regulations rather than technical faults. It has been estimated that a blade shearing off at top rotor speed from a 38 m rotor rated at 200 kW could travel up to 165 m before hitting the ground (Taylor, 1983). Safety regulations in some countries require large turbines to be at least 200 m away from houses or roads. Modern wind turbines are equipped with sophisticated monitoring and shutdown systems to limit the hazard posed by any component failure.

It may also be necessary to consider whether the proposed site for the wind turbines is of any *archaeological significance*. It is surprising how often upland sites have been inhabited by previous generations and care is needed to ensure the turbines are sited away from any sites of historical or archaeological interest.

5.3 ENVIRONMENTAL ASSESSMENT

An application by a wind turbine developer for planning authorisation or permission may include a legal requirement to carry out an environmental assessment and submit an environmental statement. In any event, preparation of such a document is good practice to ensure the project has been well planned. A comprehensive environmental impact statement is a most useful document which can be used during the period of consultation with local inhabitants. The environmental assessment covers the issues in the preceding section, requiring the developer to show that the proposed wind development conforms to various requirements. These requirements relate to such questions as visual impact, noise, effect on the local ecology and safety and the developer must show that all reasonable measures have been taken to limit adverse environmental impact.

Suggested topics which should be included in the environmental statement are listed by the British Wind Energy Association (1994).

- *Policy Framework*: how the proposed project complies with regional or national development plans
- *Site Selection*: why the particular site has been selected
- *Designations*: whether the project affects any areas designated to be of particular value, e.g. national parks
- *Visual and landscape assessment*: a map is prepared showing those areas from which the wind turbines may be seen (known as zones of visual influence); more sophisticated techniques of visual assessment may be appropriate for larger projects
- *Noise assessment*: to ensure the wind farm will not create a nuisance at local dwellings
- *Ecological assessment*: impact on flora and fauna
- *Archeological and historical assessment*: so that any locations of particular historical significance within the wind farm site can be identified and avoided
- *Hydrological assessment*: the effect of the proposed project on water courses
- *Interference with telecommunications systems*: it is especially important to ensure that there is no interference with local TV reception or disruption of microwave routes.
- *Safety assessment*: safety of the site, including structural integrity of the turbines
- *Aircraft safety*: the civil and military authorities need to be consulted to ensure there is no hazard to low flying aircraft or to radar systems

- *Traffic management and construction*: includes any modifications required to public roads
- *Electrical connection*; the environmental impact of any electrical equipment needs to be considered, e.g. overhead lines and substations
- *Economic effects on the local economy*: includes an estimate of the number of permanent and temporary jobs which may be created
- *Global environmental effects*: the global benefit of the project can be demonstrated; this is the benefit obtained by reducing emissions from fossil fuel power stations.
- *Tourism and recreational effects*: the effect of the project on tourism and recreational activities
- *Decommissioning*: proposals for the project at the end of its useful life
- *Mitigating measures*: ways in which any adverse environmental impact may be minimised
- *Non-technical summary*: a most important section to ensure the results of the environmental assessment are easily accessible to all readers

For a large scheme, an environmental assessment can be an expensive process. Extensive studies, using virtual reality or other procedures, may be carried out to determine the visual impact of the project. Computer modelling of the geography of the area may be used to show the zones of visual influence. Such computer-generated maps could also plot the relationship of the wind farm to telecommunication links, in order to consider the possibility of interference.

Figures 5.4–5.7 are photographs of existing wind farms showing how wind turbines can be integrated into the landscape after all the planning processes have been successfully completed. However, even small schemes will benefit from a formal environmental assessment and this can often be performed without the expensive professional assistance required for large projects. The use of sophisticated computer-based tools allows very detailed calculations to be carried out and the results to be presented in a most attractive manner. However, the basic steps of a sound environmental assessment do not require expensive tools; local resources are normally quite adequate.

5.4 SOCIAL CONSIDERATIONS

The acceptability of wind plant depends to a very considerable extent on public recognition of the need to conserve ever scarcer reserves of fuel and to reduce harmful emissions. An educational programme may therefore be needed to create the most favourable opinion of wind energy as part of the wider promotion of alternative energy sources.

Figure 5.4. Reproduced by permission of National Wind Power

Figure 5.5. Reproduced by permission of National Wind Power

Figure 5.6. Reproduced by permission of National Wind Power

Figure 5.7. Bryn Titli wind farm in Wales. The turbines are Bonus 450 kW machines

Figure 5.8. 300 kW wind turbine at a utility test site

An important factor here is likely to be the devolution of the responsibility for wind developments to the most local level possible, with the aim of encouraging the feeling that wind developments are owned by the local people. This is particularly important in the avoidance of the 'not in my back yard' or NIMBY syndrome.

It is interesting to note that, in the UK at least, local public support tends to increase once a wind farm has been constructed. Initial fears that the environmental impact will be severe prove groundless once the installation is seen in operation. However, a core of objectors may remain who do not approve of the wind farm.

SUMMARY OF THE CHAPTER

Planning procedures in an industrialised country may be more rigorous because of established planning regulations, but the impact of wind power development on existing energy supplies and on the environment has to be taken into account by any society. The main benefit of wind power is that it is a renewable resource which reduces overall levels of harmful emissions by displacing the need for fossil fuel power plants. The public's perception of wind power is more subjective than its attitude towards many other industries because of the subjective nature of problems like visual impact and noise.

The active land area required for wind turbines is very small and the ground between machines is not lost to agriculture. Turbine noise can be overcome by good design. Wind turbines can have a detrimental effect on bird life, although in most cases this appears to be small, and the sensitive siting of machines is important to avoid migratory routes and nesting areas. The public acceptance of wind energy projects will depend on attitudes towards the need for renewable energy as well as recognition of well planned and efficiently operated wind plants.

BIBLIOGRAPHY AND REFERENCES

Benner, J.H.B., Impact of Wind Turbines on Bird Life, in *Proceedings of the European Community Wind Energy Conference*, Lubeck-Travemunde, Germany, 8–12 March 1993, pp. 20–23.

British Wind Energy Association, *Best Practice Guidelines for Wind Energy Development*, BWEA 1994, ISBN 1 870064 21 6.

Clausager I., Impact of wind turbines on birds – an overview of European and American experience, in *Birds and wind turbines: can they co-exist*? ETSU Publication ETSU-N-133, 1996.

Department of Trade and Industry *Renewable Energy (with an annex on wind power)*. Planning Policy Guideline Note PPG22, HMSO, London, 1993.

Garrad, A.D., *Time for Action: Wind Energy in Europe*, European Wind Energy Association, Rome, 1991.

Hagg, F. Bruggeman, J.C. and Dassen, A.G.M. *et al.* National Aero-Acoustic Research on Wind Turbines in the Netherlands, in *Proceedings of the European Community Wind Energy Conference*, Lübeck-Travemunde, Germany, 8–12 March 1993, pp. 290–293.

International Energy Agency, *Recommended Practices for Wind Turbine Testing and Evaluation*, Vol. 4, *Acoustics: Measurement of Noise Emission from Wind Turbines*, 3rd ed., edited by S. Ljunggren, 1994.

Taylor, R.H., *Alternative Energy Sources for the Centralised Generation of Electricity*, Adam Hilger, Bristol, UK, 1983.

Wagner S. Bareis, R. and Guidati, G. *Wind Turbine Noise*, Springer-Verlag, Berlin, 1996.

World Energy Council, *New Renewable Energy Sources*, Kogan Page, London, 1994.

SELF-ASSESSMENT QUESTIONS

Part A. True or False?

1. Rigorous planning laws relating to wind turbine developments are most likely in less developed, sparsely populated countries.

2. The attitude of the general public to renewable energy in general is an important factor relating to planning authorisation for wind turbines.

3. The energy used during its manufacture is more than a wind turbine will generate in its lifetime.

4. Wind turbines help to mitigate the greenhouse effect.

5. Public reaction to the appearance of wind turbines is always favourable.

6. The total land area occupied by a wind farm cannot be used for any other purpose.

7. Wind turbines have no effect on local ecology.

8. At the planning stage, wind turbine noise must be taken into account.

9. Wind turbines can cause radio and TV interference.

10. Wind turbines have a poor safety record.

PART B

1. Which factors influencing planning for wind power are most likely to be shaped by public attitudes?

2. Name two kinds of costs which should be taken into account when planning for wind power.

3. Name some environmental benefits from wind power.

4. Name some environmental disadvantages of wind power.

5. What is the best way to ensure the public acceptability of wind turbine projects?

Answers

Part A

1. False; 2. True; 3. False; 4. True; 5. False; 6. False; 7. False; 8. True; 9. True; 10. False;

Part B

1. The attitude of the general public to wind power will be reflected in the development of the planning policy within the national planning system. This in turn will influence the way guidelines will be written on how planning permission or authorisation should be obtained by developers.

2. Costs imposed by planning requirements and costs of integrating wind power into electricity networks need to be taken into account. For example, planners may require wind turbines to be spaced further away from each other or from habitations, thus increasing wind farm costs. Such requirements could also increase the cost of connection into the electricity network.

3. Wind power reduces emissions of pollutants and conserves irreplaceable resources because it cut downs the use of fossil fuels.

4. Visual intrusion, land use, impact on local ecology, noise and interference with telecommunications are all disadvantages of wind power; they can be mitigated by careful planning.

5. Wind turbine projects are most likely to be acceptable where local people are involved in decision making.

6

Utility-Connected Wind Turbines

AIMS

This unit discusses the factors which should be taken into account when wind turbines are used to supply electricity to utility networks. The design of wind farms is also discussed.

OBJECTIVES

When you have completed this unit you will have an appreciation of four areas:

1. The physical and institutional structure of an electric utility power system.
2. Factors concerned with integration of wind energy into utility networks.
3 The impact of wind power on central generators, transmission networks and distribution networks.
4 The design of wind farms.

NOTATION AND UNITS

	Symbol	Units
D	Rotor diameter (used to indicate turbine spacing)	–
P	Real power	W, kW
Q	Reactive power	VAr, kVAr
R	Resistance	ohm
V	Voltage	volt
X	Inductive reactance	ohm
W	Losses	W, kW

6.1 ELECTRIC POWER SYSTEMS

In the early days of electricity supply each town or city had its own generating station supplying a restricted area. The size of the generators was small and, in each station, at least one generator had to be kept running as a reserve in case of breakdowns. Then from about 1930 it was found to be cost-effective to interconnect these individual power stations so the total number of generators held in reserve could be minimised and larger, more efficient generating units could be constructed. This led to the development of a highly interconnected *transmission network or grid* in most countries and the construction of very large, *central* generating stations. Power was then generated by these large central generating stations, fed up to the grid and transmitted long distances to be transformed down to the distribution systems then finally delivered to the consumers. This arrangement is shown schematically in Figure 6.1

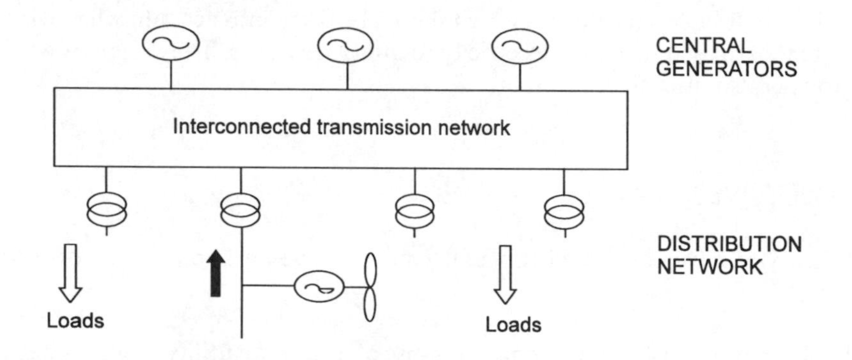

Figure 6.1. A typical large utility power system

Some of the transmission networks are extremely large, e.g. one grid covers most of western Europe and each power station may be of up to 2000 MW in capacity. In some countries this centralised approach was limited by the additional requirement to provide heat from the power stations and, as heat cannot easily be transported long distances, some small local power stations were retained. However, the most usual arrangement for a power system is that shown in Figure 6.1.

Wind energy is obviously a diffuse energy source, so wind farms must be distributed over a wide area. The size of wind power installations may vary from individual wind turbines at say 25 kW to large wind farms of up to 30 MW. Therefore, wind turbines and wind farms are connected not to the transmission grid, but to the distribution circuits which supply customers. Thus, wind farms are *embedded* in the distribution system. The use of generators embedded into the distribution circuits is something of a return to the early days of electricity supply when local generating stations supplied local loads, although now there is also a connection to the main transmission grid. Distribution systems that include *embedded generation* are rather more complex to design and operate than those which simply take power from the transmission network and deliver it to consumers.

A further, non-technical, issue which has emerged recently has been the changes in structure and ownership of the electricity supply system in many countries. Until a few years ago, it was normal for a single state-owned utility to own and operate the central generating stations, the transmission grid and sometimes the distribution systems. Decisions on investment and operation policy were then made by the utility after an internal cost-benefit analysis. This is still the case in France where a single utility Electricité de France (EDF) has a monopoly on almost all aspects of electricity supply. But an alternative structure has recently emerged in several countries. The generating stations have been sold off to private companies who now offer to sell electricity to the distribution companies through a market mechanism. The transmission system is then operated by an independent company as a power transportation business. This market-based structure occurs in Chile, the UK and parts of Scandinavia. Although the basic requirements of generating and delivering power have not changed, the new arrangements probably offer more scope for the independent generation of electricity from a variety of sources, including wind power, provided they can compete in the market environment.

6.2 INTEGRATION ISSUES

Integration of wind power into an electric power system requires consideration of several topics:

For the central generators

- Energy credit
- Capacity credit
- Control issues

For the transmission grid

- Operating costs
- Requirement for equipment

For the distribution system

- Operating costs
- Requirement for equipment
- Power quality
- Protection and safety

Whether wind energy is beneficial in each aspect depends on a number of factors, some related to the wind plant:

- Wind potential, predictability and performance
- The accuracy of load forecasts, load variation and load management
- Types of wind turbines and wind farms
- The mix of generating plant in the utility system
- Variable costs of the utility such as the cost of fuel and some maintenance costs
- Fixed costs of the utility such as capital and other investment-related costs

None of these factors can be considered in isolation, so an approach is required in which economic and technical aspects of the utility systems are modelled as a whole, firstly without wind and secondly with wind capacity. A fundamental requirement is that the quality of the electricity supply to the consumer should be unchanged by the introduction of wind power capacity. These topics are discussed more fully in the following sections.

6.3 IMPACT ON CENTRAL GENERATION

Like other generating plant, wind power has fixed (unavoidable) costs, consisting of capital and related costs which do not vary with electricity produced (e.g. the purchase cost of the wind turbines). There are also variable (avoidable) costs, which are smaller than those for conventional generating plant because no fuel is burnt (e.g. some maintenance costs). The total costs of wind power are relatively straightforward to determine, with the proviso that as yet there is no really long-term experience on life-dependent costs.

The savings result from savings in fuel and other variable costs, and in the longer term from the need for less conventional capacity. However, these savings are offset by the need for increased spinning reserve (fossil fuel plants which are kept on standby for immediate progression to full output) and the costs resulting from greater load variations on conventional plant.

For the purposes of these studies, the utility cost savings are divided into energy and capacity credits, in which all variable and fixed savings are included.

Energy credit

Fuel savings, or energy credits, are the major benefit from wind power. These savings result from the reduced need to run other generating plant. This in turn results in lower fuel and related variable costs, including maintenance and staff costs.

The expected annual energy output as a function of the annual wind velocity and the main data for typical wind turbine sizes are given in Table 6.1. The frequency distribution of the wind velocity is assumed to be a Weibull distribution with a shape factor k equal to 2. For large power systems in which there is only a small penetration of wind turbines the energy output of a wind farm is approximately equal to the energy not required from conventional plant. Thus each kWh generated by the wind turbines results in one kWh of output saved from conventional generation. However, in the future when wind energy starts to become a significant proportion of the generation capacity (e.g. over 5%) the effect of wind energy on the variation of the load seen by conventional plant will become important and fuel savings will be reduced.

Table 6.1. Examples of annual output (MWh/year) for different sizes of grid connected turbines: the figures are based on measured power curves

	Wind turbine rated output and main dimensions		
Annual mean wind velocity at hub height (m/s)	100 kW Hub height 24 m Rotor dia. 20 m	200 kW Hub height 30 m Rotor dia. 25 m	400 kW Hub height 30 m Rotor dia. 35 m
4	65	120	180
5	125	240	415
6	195	380	715
7	265	535	1025
8	330	685	1325
9	390	820	1585
10	435	935	1805

Environmental damage is an important variable cost of conventional plant and, although often difficult to calculate, this is likely to be of significant benefit for wind power.

Capacity credit

Wind plant can only earn capacity credit if it defers the installation of new generating plant and thus defers capital expenditure. There are, in essence, two main methods of calculating capacity credit:

1. The contribution made by wind plant at times of peak demand on a utility is assessed over a period of years and the average power at these times is taken as the capacity credit.

2. The loss of load probability (LOLP) for a utility network is calculated, initially in the absence of wind generation (Billinton & Allan, 1984). It is then recalculated with wind generation on the system and conventional

plant notionally removed until the initial level of LOLP is obtained. The power of this subtracted plant is the capacity credit of wind.

Both methods yield very similar results. For small penetrations of wind power, the capacity credit is usually close to the average output of the wind plant, i.e. 1000 MW of wind plant generally has a capacity credit of about 300 MW.

Control issues

Wind capacity imposes additional control needs on a utility system. At low penetrations wind power tends only to reduce the demand on the central generators. However, with high penetrations, because of the stochastic nature of the wind, as the mean value of the demand is decreased the uncertainty or variance of the demand will increase. This is further complicated by the difficulty in predicting the output of wind turbines, say 24 hours ahead. This prediction is made routinely by the utilities for customer loads. The utilities are faced with a greater uncertainty in the load that must be met from the central generating plant. Some types of generators (e.g. hydro sets) are able to follow the demand easily and with little penalty for running at reduced output. However, large steam generators are not able to follow the load demand rapidly, and due to their lower thermal efficiency, they incur significant additional costs when operating at part load.

The frequency of a power system (nominally either 50 or 60 Hz) is determined by the balance between the load and generation. If the load exceeds the output power of the generators, they will slow down and the frequency will fall. Conversely, if the load is smaller than the generator output, the frequency will rise. Adding wind generation and therefore increasing the variability of the load seen by the central generators will have some effect on the frequency. However, in large power systems this effect will be so small it can be neglected for practical purposes.

6.4 IMPACT ON THE TRANSMISSION SYSTEM

Generation embedded in the distribution system reduces the power required to flow through the transmission system as some of the consumer load is now met from local generation. This has two effects on the transmission system: lower losses in the transmission system and reduced requirements for plant, e.g. lines and transformers. But care should be exercised when evaluating these benefits; although the mean value of the load may be reduced, the variation of the load passing through the transmission network will be increased.

6.5 IMPACT ON THE DISTRIBUTION SYSTEM

The impact on the distribution system (Jenkins, 1996) is similar to the impact on the transmission system, i.e. a change in network losses and the requirement for equipment (e.g. lines and transformers). However, as the distribution network is sparser and has fewer interconnecting circuits, there is less scope for reducing the requirement for lines and transformers. On some days the wind will not blow, so the circuits must be available to meet the load on those occasions. Circuits are also required to connect the wind farms to the transmission system and to supply reactive power.

Distribution systems are used to supply consumers with electricity and to ensure the voltage of the circuits is held within close limits. When connecting wind turbines it is important to ensure the additional flows of real and reactive power do not cause the limits on voltage to be exceeded. Figure 6.2 shows a simple representation of a fixed-speed induction machine with only limited power factor correction connected to a radial distribution circuit. The wind turbine exports real power, P, and imports reactive power, Q. The radial distribution circuit is represented by a simple series impedance of resistance R and inductive reactance X. The bus-bar V_0 represents the connection to the transmission system; it is called an infinite bus-bar because its voltage is assumed to be constant under all conditions.

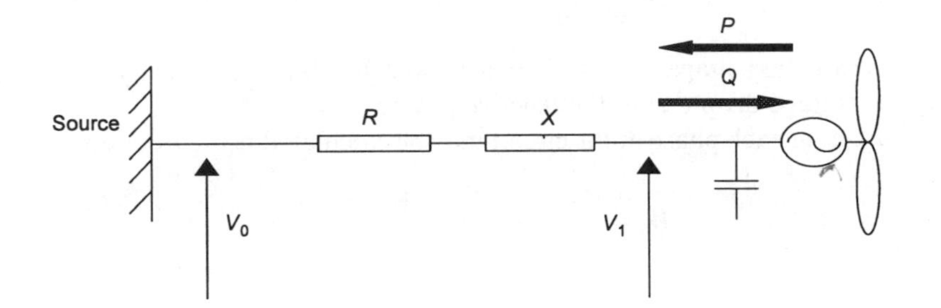

Figure 6.2. Fixed speed wind turbine on a radial circuit

The voltage at the wind turbine may be calculated approximately by

$$V_1 = V_0 + \frac{PR - QX}{V_0} \tag{6.1}$$

This simple expression shows that the real power, P, passing through the network resistance, R, causes a rise in the voltage at the wind turbine, whereas the reactive power, Q, with the inductive reactance of the circuit, X, cause a lowering of the voltage.

The electrical losses in the circuit may be calculated from:

$$W = \frac{(P^2 + Q^2)R}{V_0^2} \tag{6.2}$$

A simple example will illustrate these calculations. Consider a 600 kW wind turbine operating at rated output and absorbing 210 kVAr. It is connected to a 415 $V_{(line-line)}$ system and the impedance of the connection when referred to 415 volts is

$$R = 0.007 \text{ ohm}$$
$$X = 0.014 \text{ ohm}$$

Now both equations (6.1) and (6.2) apply to each phase of the three-phase network, so for each phase

$$P = 200 \text{ kW}$$
$$Q = 70 \text{ kVAr}$$
$$V = 240 \text{ V } (415/\sqrt{3})$$

Then

$$V_1 = 240 + \frac{(200 \times 10^3 \times 0.007 - 70 \times 10^3 \times 0.014)}{240}$$
$$V_1 = 240 + 5.8 - 4.1$$
$$V_1 = 242 \text{ V}_{\text{phase}}$$
$$V_1 = 419 \text{ V}_{(\text{line-line})}$$

In this case the voltage rise due the real power flow is almost exactly cancelled by the voltage drop due to the reactive power flow.
The losses in each phase of the circuit may be calculated using Equation (6.2):

$$W = \frac{(200^2 + 70^2) \times 10^6 \times 0.007}{240^2}$$
$$W = 5456 \text{ W}$$

so the total losses in the circuit are 16.4 kW. In this simple example the voltage rise at the wind turbine is small and the electrical losses modest. However, this may not always be the case, then more sophisticated calculation techniques are required to give more precise predictions of the behaviour of the circuit.

In addition to steady-state voltage variations, the *dynamic voltage variations* caused by wind turbines may lead to restrictions on their use. These dynamic voltage variations are often called *flicker* because of the effect they have on incandescent lights. The human eye is very sensitive to changes in light intensity, particularly if the variation occurs at frequencies around 10 Hz. The torque from a horizontal-axis wind turbine rotor contains a periodic component at the frequency the blades pass the tower. This cyclic torque is due to the variation in wind speed seen by the blade as it rotates. The variation in wind speed is due to a combination of tower shadow, wind shear and turbulence. The rotor torque variation is then translated into a change in the output power and hence a voltage variation on the network.

For a large wind turbine the blade passing frequency will be around 1–2 Hz and, although the eye is less sensitive at this frequency, it will still detect voltage variations greater than about 0.5%. In general the torque fluctuations of individual wind turbines in a wind farm are not synchronised, so the effect in large wind farms is reduced as the variations average out. However, it has sometimes been noted that the wind turbine rotors in a wind farm fall into synchronism, then significant power and voltage fluctuations can be measured on the network. These dynamic voltage variations are sometimes known as issues of *power quality*, although this term also includes harmonic distortion of the network voltage. Harmonic distortion is mainly of concern with variable-speed wind turbines where the power electronic converters may produce output currents that are not perfect sine waves.

The power factor of a fixed-speed wind turbine may be improved by installing capacitors in the base of the tower. If enough capacitors are fitted to supply completely the no-load reactive power requirement of the wind turbine, there is the possibility of *self-excitation* and potentially hazardous overvoltages may be generated. Self-excitation may occur if the wind turbine and capacitors are disconnected from the network. In this case stability is lost and the capacitor bank may resonate with the inductance of the generator. This resonant condition may produce high voltages, possibly hazardous to any equipment connected to the part of the network that has become isolated with the turbine. Self-excitation may be avoided by limiting the power factor correction to a value that will not supply the reactive demand of the generator, even when the frequency of the isolated circuit rises.

Most utilities are extremely sensitive to the operation of wind turbines on isolated or *islanded* sections of the network. The reasons include the possibility of supplying consumers with power that is outside the normal limits for frequency and voltage, operating systems without a proper earth and continuing to supply power into a fault. Therefore it is usual for each wind turbine, and the wind farm, to have special protective relays to detect this *islanded* condition and to shut down the turbines.

The impedance of the network from the point of connection to the infinite bus-bar, represented by R and X in Figure 6.2, also determines the *fault level*. The *fault level* or short-circuit level, S, is usually expressed in MVA and is an indication of the 'strength' of the network. If a three-phase short circuit occurs at the point of connection, the impedance of the network back to the infinite bus-bar determines the magnitude of the fault current that will flow. A low source impedance will allow high fault currents to flow, hence it leads to a high fault level.

The magnitude of the source impedance is the inverse of the fault level once they are both expressed in compatible units. In some countries it is accepted practice to define the size of wind turbine or wind farms which may be connected at a given point in the network in terms of a ratio of the wind farm rating (MW) to the network fault level (MVA). A conservative ratio is 4% and adopting this value means that problems with voltage disturbance are unlikely. However, in many situations it is not possible to obtain a ratio of 4%, then more detailed calculations are required.

Fixed-speed wind turbines themselves will contribute to the fault level of the distribution network. This is unlikely to be a problem in rural areas where the network is weak and the existing fault levels low. However, if wind farms are installed near to industrial plants or where the electrical network is strong, an increase in the fault level may create difficulties. Electrical plant (e.g. circuit-breakers) is designed to operate satisfactorily with a defined fault level; if this fault level is exceeded, the equipment may become overstressed and fail.

6.6 WIND FARMS

Large-scale utilisation of wind energy at a site is obtained by installing several turbines in clusters or wind farms. This has an economic advantage as well as a benefit in connection with operation and maintenance of the plant. Any wind energy project will incur fixed costs such as the preparation of the environmental statement, legal fees and project management costs. They are largely independent of the size of the wind farm, so it is desirable to spread them over as large a project as possible. This has led to quite large wind farms being constructed in Europe, some of them having as many as 100 turbines and an output of 30 MW. And wind farms with even more turbines are not uncommon in the United States.

By installing several wind turbines in a cluster, each turbine will generate a wake that will affect the output from the other turbines in the farm. The influence of a turbine on the wind velocity profile is illustrated in Figure 6.3. Various theories for estimation of wake effects have been developed and full-scale experiments have been conducted in order to verify them. Due to the complexity of the phenomena, no final solution has been found. However, a simple method for calculation of the wake effects is illustrated in Figure 6.4. Using the balance in momentum and assuming the wind velocity behind the first rotor disc is one-third of the upstream wind velocity, an expression for the wind velocity behind the rotor can be derived (Jensen, 1983):

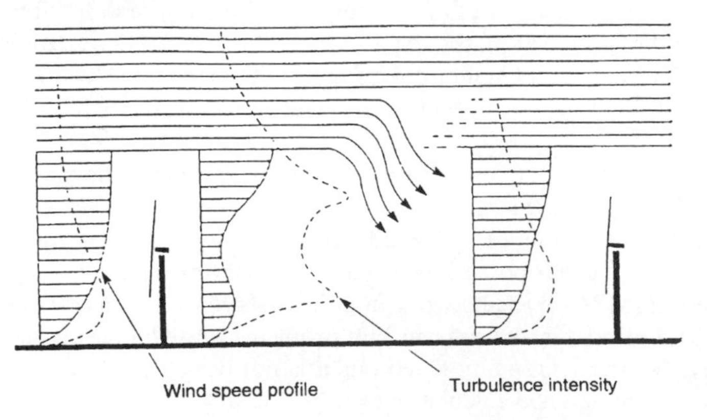

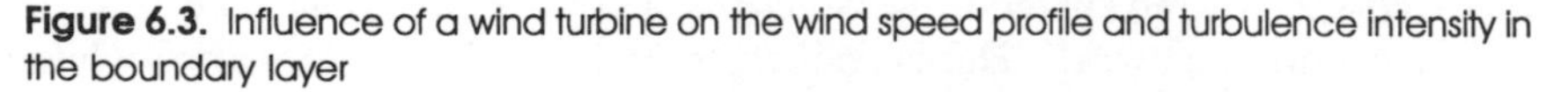

Figure 6.3. Influence of a wind turbine on the wind speed profile and turbulence intensity in the boundary layer

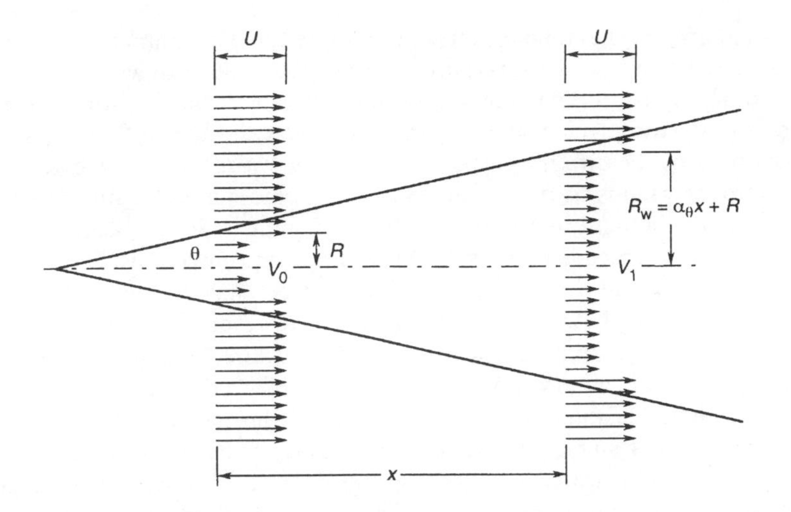

Figure 6.4. Parameters for the simple wake model of Jensen (1983)

$$V = u\left(1 - \frac{2}{3}\left(\frac{R}{R + \alpha_\theta x}\right)^2\right) \tag{6.3}$$

where V is the wind velocity behind the rotor, u is the wind velocity in front of the rotor, R is the rotor radius, α_θ is a constant specifying the wake typically 0.1 (Figure 6.4), and x is the distance between the turbines.
The influence on one turbine from several turbines is calculated by adding the wakes from the individual turbines.
The influence of the wake effects on the total output from a wind farm is expressed by the array efficiency, which is given by

$$\frac{\text{Annual energy of whole array}}{\text{Annual energy of single turbine} \times \text{total number of turbines}} \tag{6.4}$$

The array efficiencies for some simple wind farm configurations are given in Table 6.2.

Table 6.2. Example of array efficiency for different wind farm configurations calculated according to the model given by Jensen (1983)

Number of rows	Turbine spacing in rotor diameters (D) Along row	Between rows	Average annual array efficiency %
1	3D		91
2	3D	3D	79
3	3D	7D	84
4	3D	7D	93
5	3D	5D	84

The wind around the wake of a turbine is more turbulent than the upstream wind, so a further consideration when deciding the location of wind turbines is to ensure they do not run in the wake of an upstream turbine for long periods. Running in turbulent wake is likely to increase the fatigue damage experienced by the turbine, particularly on the yaw drive and yaw bearing.

Where the wind is strongly directional (e.g. some sites in California where the winds are thermally driven) it may be appropriate to locate the turbines in rows that are perpendicular to the prevailing wind direction. The spacing between turbines in a row is then reduced to perhaps $2D$ or $3D$ but with significant spacing between rows of say $10D$. However, on many sites there will not be a single well-defined prevailing wind direction, so a more even spacing of 5–$7D$ in all directions may be adopted. The denser the wind farm, the lower the cable and road costs. Set against this is the reduction in output due to array losses and the increase in damage to the turbines because of turbulence. Sophisticated computer programs are now available for the *micrositing* of wind turbines. When calculating the output of the wind farm, these programs take account of the local topography and the interaction with adjacent turbines.

In addition to array losses, it is also necessary to calculate the electrical losses within the wind farm as they will reduce the power exported to the network. Figure 6.5 shows the typical arrangement for the collection of power from a wind farm. There is a transformer adjacent to each turbine and underground cables bringing the power to a larger transformer for export to the public electricity network. The main losses will be within the transformers, although there will also be smaller losses in the cables. There will also be some power consumed by the turbine control circuitry when the wind speeds are too low for the turbines to operate. On a well-designed wind

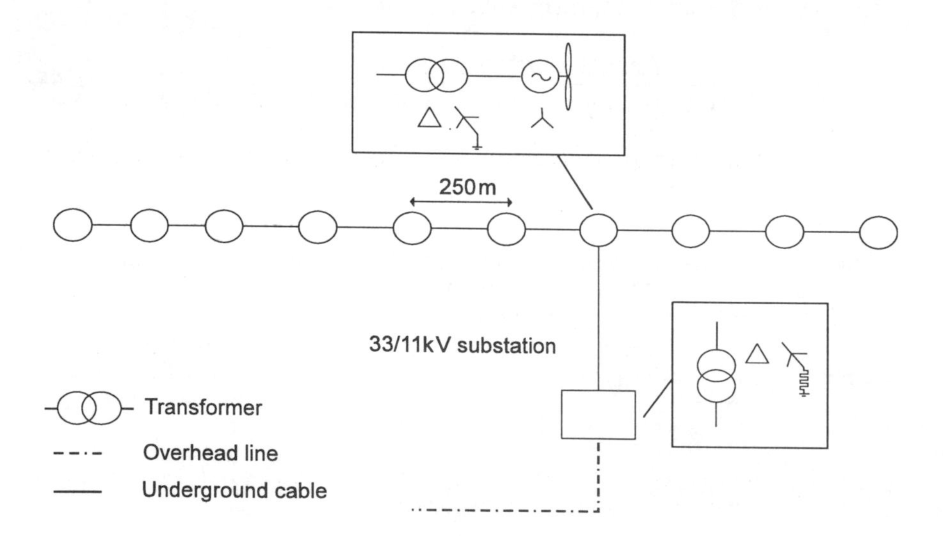

Figure 6.5. Windfarm layout (showing transformers and earthing arrangements)

farm the internal electrical losses should be only a few percent of the total output but this can only be achieved if the transformers and cables are sized carefully.

The utility is likely to require sophisticated electrical protection at the point of connection of the wind farm to the network. This is to fulfil three requirements:

- To isolate the wind farm in case of any faults on the internal electrical system or turbines
- To stop the wind farm feeding power into any fault on the utility network
- To shut down the wind farm if any adjacent section of public network becomes islanded

SUMMARY OF THE CHAPTER

The four factors that affect the integration of wind power into utility network were examined: energy credit, capacity credit, control problems and electrical considerations. An approach was described in which the economic and technical aspects of the system, with and without the wind power component, were modelled as a whole. The need to maintain the grid power quality was emphasised. The problem of wake interactions from individual turbines in wind farms was considered and an empirical method to allow for the overall array efficiency was described.

BIBLIOGRAPHY AND REFERENCES

BILLINTON, R., AND ALLEN, R N., *Reliability evaluation of power systems* Plenum Publishing, New York 1984.

JENKINS N., Embedded generation tutorial: 1 and 2, *IEE Power Engineering Journal*, June 1995 and October 1996.

Jensen, N.O., *A Note on Wind Generator Interaction*, Risø National Laboratory, Denmark, 1983.

SELF-ASSESSMENT QUESTIONS

PART A. True or False?

1. The introduction of wind power capacity to a utility grid should not change the quality of the electricity supply to the consumer.

2. Capital costs of power plant vary with the quantity of electricity produced.

3. If wind energy production is high the wind energy credits are not affected.

4. Thermal power units are not affected by wind power when considering the control needs of a utility system.

5. Wind turbines will affect distribution network voltages.

6. Wind turbines will affect network frequency.

7. Wakes from wind turbines enhance the efficiency of a wind farm.

8. Self-excitation of induction generators is desirable.

9. Fixed-speed wind turbines will increase the network fault level.

10. Harmonics and flicker improve the quality of the voltage on the public electricity network.

PART B

1. What is the annual energy output from a wind farm consisting of 16 turbines with a rated power of 200 kW, hub height 30 m and rotor diameter 25 m? The turbines are in two rows 7 diameters apart and the distance between turbines is 3 rotor diameters. The wind speed at hub height is 8 m/s and the shape factor of the distribution is 2.0.

2. An 800 kW wind turbine is generating at rated power and absorbing 300 kVAr. It is connected to a 690 $V_{(line–line)}$ three-phase circuit which has a source impedance of $R = 0.02$ ohm and $X = 0.04$ ohm (both values referred to 690 V). Calculate the change in voltage at the point of connection caused by the operation of the turbine.

3. Calculate the electrical losses in the connecting circuit of Question 2.

4. The size of the local power factor correction capacitors of the wind turbine described in Question 2 are increased so there is no reactive power drawn by the wind turbine. What is the voltage at the point of connection with the turbine generating 600 kW?

5. What is likely to occur if the wind turbine and capacitors are disconnected from the network in strong winds?

Answers

Part A

1. True; 2. False; 3. False; 4. False; 5. True; 6. False; 7. False; 8. False; 9 True; 10. False.

Part B

1. From Table 6.1 the output of a single turbine is 685 MWh/year and from Table 6.2 the array efficiency is 0.84. Hence the wind farm output = 685 × 16 × 0.84 = 9206 MWh/year.

2.

$$V_0 = \frac{690}{\sqrt{3}} + \frac{(266.6 \times 0.02 - 100 \times 0.04) \times 10^3}{398}$$
$$V_0 = 401\,\mathrm{V_{phase}}$$
$$V_0 = 696\,\mathrm{V_{(line-line)}}$$

Therefore the change in voltage is a rise of 6 V.

3.

$$W = \left[\frac{(266.6^2 + 100^2) \times 10^6 \times 0.02}{398^2}\right] \times 3$$
$$W = 30.7\,\mathrm{kW}$$

Therefore the electrical losses in the connecting circuit are 31 kW.

4.

$$V_0 = \frac{690}{\sqrt{3}} + \frac{266.6 \times 10^3 \times 0.02}{398}$$
$$V_0 = 411\,\mathrm{V_{phase}}$$
$$V_0 = 713\,\mathrm{V_{(line-line)}}$$

By eliminating the requirement for reactive power, the rise in voltage at the point of connection has increased to 23 V.

5. If the wind turbine and capacitors are disconnected from the network, the wind turbine will accelerate as it is no longer supplying any load. The capacitors will then resonate with the generator and very high overvoltages can occur very quickly.

7

Applications

AIMS

This unit discusses the factors which should be taken into account when wind turbines are used in a variety of applications other than the generation of electricity into large, land-based utility networks. Combined wind–diesel systems, water pumping and offshore siting are important applications considered here.

OBJECTIVES

When you have completed this unit you will have an appreciation of three areas:

1. Design and performance of wind–diesel systems.
2. Design and performance of wind turbines for water pumping.
3. Factors concerned with the offshore siting of wind turbines.

NOTATION AND UNITS

Symbol		Units
A	Area	m^2
C_p	Coefficient of performance or power coefficient, a measure of rotor efficiency	
D	Rotor diameter	m
g	Acceleration due to gravity	m/s^2
H	Head	m
J	Moment of inertia	$kg\,m^2$
P	Power	W
Q	Volume flow rate	m^3/s

Symbol		Units
T	Torque	N m
V	Wind speed	m/s
ω	Rotational speed	rad/s
ρ	Density of air	kg/m^3
ρ_w	Density of water	kg/m^3

7.1 WIND–DIESEL SYSTEMS

7.1.1 System Configuration

In many remote and sparsely populated areas of the world there is no transmission grid as it is too difficult and expensive to build the circuits to supply only a few consumers. Hence diesel generator sets may be the only method of providing electricity. However, operating diesel generator sets in remote areas is not straightforward. The difficulties include availability of fuel, spare parts and skilled labour for maintenance as well as the challenging problem of meeting the variable load of a small community in a cost-effective manner.

Wind and diesel power plant can be used in combination for power supply of autonomous system where a connection to the national grid is either impossible or too expensive due to long transmission lines. The diesel generator(s) provide power when needed, using the energy stored in the diesel fuel, whereas the wind turbine(s) act to reduce the diesel fuel consumption and the operating time of the diesel engine(s). Wind–diesel systems to provide power in remote areas of developing countries have significant social benefits, but their cost-effective and reliable operation has proven difficult. Recent progress is beginning to tackle the problem.

In principle the power supply system required for a small isolated village is no different from that needed for a major industrialised country. The village system may be many orders of magnitude smaller, but if an alternating current system is used, it is still necessary to supply the consumers with electricity at a controlled voltage and frequency. This is the same challenge as faces major power utilities operating national power systems, although in the small system, the acceptable limits on voltage and frequency may be rather wider. The small village power system may actually be more difficult to control because the small size of equipment means that changes in frequency will occur faster.

As discussed in Chapter 6, the frequency of a power system is determined by the balance between input power from the generators and the power taken from the load. If the generator output exceeds the load, the surplus power will act to increase the rotational speed of the generators and hence the system frequency. Conversely, if the load demand exceeds the power available from the generator, the frequency will fall. In a major power system the inertia of the generators is very large, so the effect of the power generated from wind turbines has only a very small effect on the system frequency and this can usually be ignored. However, in a wind–diesel system the inertia of the generators is small, so any surplus power will produce a rapid rise in the rotational speed of the generators and therefore in the system frequency. This variation in rotational speed can be expressed using the simple dynamic equation of the rotating masses of the generators:

$$
\begin{aligned}
J\frac{\partial \omega}{\partial t} &= T_{gen} - T_{load} \\
\omega J\frac{\partial \omega}{\partial t} &= P_{gen} - P_{load} \\
\frac{\partial \omega}{\partial t} &= \frac{1}{\omega J}(P_{gen} - P_{load})
\end{aligned}
\tag{7.1}
$$

where

J = the total moment of inertia of all the spinning masses on the system (i.e. generators and any rotating loads)

T_{gen} = the torque applied by the generators

T_{load} = the equivalent torque absorbed by the loads

P_{gen} = the power supplied by the generators

P_{load} = the power absorbed by the loads

ω = the rotational speed of the generators; it is related to the system frequency by a constant that depends on the construction of the generators

From equation (7.1) it can be seen that control of frequency is controlled by real power flows and, to a first approximation, it has nothing to do with voltage. Therefore, if the load is determined by consumer requirements, the generators must follow the load or an energy store must be used. A wind turbine that relies on the varying wind for its energy cannot be controlled to increase its output so it follows a rise in the consumer load, and any wind diesel–system must therefore run one of the diesel generators continuously or it must have an energy store. If a large wind turbine is used in a good wind climate, it may be possible to arrange for the output to exceed the load for long periods. Then instead of an energy store, a dump load may be used to regulate the system frequency; energy stores are normally provided by the diesel fuel. But although this system is simple, it does mean that large quantities of wind energy are dumped.

Figure 7.1 shows the typical fuel consumption of a diesel generator set. The fuel consumption at no-load is significant and it is undesirable to operate diesel engines at low loads for long periods as this increases wear on the pistons and cylinders. A typical value of the permitted minimum load is 40% of the rated load, producing a minimum fuel consumption of perhaps 65% of the consumption at rated load. If the diesel engine is not shut down, the fuel

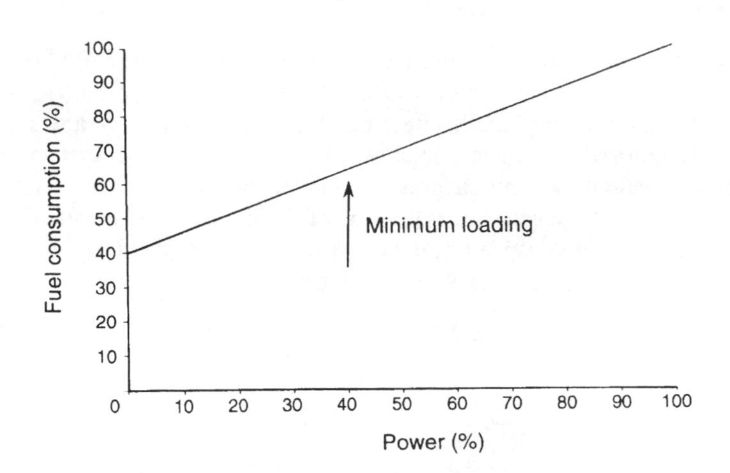

Figure 7.1. Typical fuel consumption of a diesel generator set

savings are modest, even if wind power is available to supply the load. For high fuel savings the engine must be shut down during periods when the available wind power exceeds the power required by the load. This leads to a requirement either for an electrical energy store (e.g. a flywheel or battery) or an oversized wind turbine.

Storage of electrical energy is notoriously difficult; although it is possible with batteries, flywheels and other devices, storage adds considerable expense and complexity to any wind–diesel scheme. Consequently, the control of frequency in a wind–diesel system is likely to be difficult and expensive, even though it may achieve significant savings in diesel fuel consumption. A key decision for the controller of any wind–diesel system is to decide when to start and stop the diesel engine. The engine should only be operating when the wind turbine and energy store cannot supply the load. But the variable nature of the wind and the load make it hard to predict this reliably, especially as large diesel engines take some seconds to start and they should not be started and stopped too frequently.

It is also necessary to control the voltage of the wind–diesel system. Most large wind turbines use induction generators which have no means of controlling their output voltage. In general, induction generators cannot be operated in a stable manner on their own as they need a supply of reactive power from a synchronous machine in order to generate. The control of voltage on a normal wind–diesel system is therefore carried out by varying the excitation of a synchronous generator. The excitation system on a synchronous generator varies the magnetic field of the rotor; an increase in the field raises the output voltage whereas a decrease in the field lowers the output voltage. Large wind–diesel systems, greater than say 50–100 kW, will usually consist of one or more wind turbines with induction generators and one or more diesel generators with synchronous machines. At least one synchronous generator will be connected to the system at all times to provide voltage control. The frequency is controlled by the governor of the

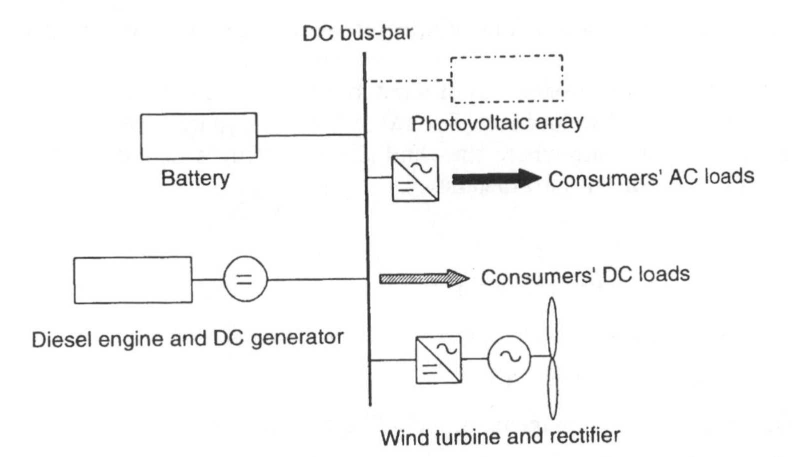

Figure 7.2. Simple DC wind–diesel system (control system not shown)

diesel engine, the dump load or sometimes by fast-acting pitch control of the wind turbines. Only one or two designs of small wind turbines have the complex pitch control mechanisms required for this duty.

Some rather smaller direct current (DC) wind diesel systems use batteries to store their energy. The output of the diesel generator and wind turbine is rectified to DC, stored in a battery then inverted to an alternating current supply (AC) using power inverters. These systems have less complex dynamic behaviour than the larger AC systems but become rather expensive as the rating increases. Figure 7.2 shows an example of a simple DC wind–diesel system and illustrates how a photovoltaic array can be easily integrated.

In a large power system it is usual to accept that the consumer load must always be met irrespective of the cost involved. However, a number of successful wind–diesel systems have been installed using a system of managing the consumer load in response to the availability of power from the wind turbine(s). The power from a wind turbine has essentially a zero marginal cost, so it should be used as much as possible. The power from the diesel generator has quite a high marginal cost (i.e. the rise in fuel use and increase in other variable costs), so it should only be used if necessary. The control system is therefore arranged so that, when power generated from the wind is available, all types of appliances may be used (e.g. lighting, space heating, water heating) at a low tariff. But when the power is generated from the diesel engine, the tariff is much higher to encourage use of only lighting and other essential loads. Some systems also use frequency-sensitive switches on consumer loads; this is to control the power absorbed by the loads, thereby regulating the system frequency.

Finally, note that the technology of high power electronics is advancing rapidly and large power converters can be used to control real power flows in and out of battery and flywheel energy stores and also to regulate voltage. In spite of these advances, there remains the fundamental limitation of all AC power systems: real power flows control frequency and reactive power flows control voltage.

The many layout possibilities of wind–diesel systems are characteristic of their early state of development. Several of the design parameters are heavily determined by the site where the wind–diesel system is installed. Here are some of the important parameters:

- Wind conditions
- Electricity consumption patterns
- Correlation between variation in wind and electricity consumption
- Power quality requirements
- Demand or load management, including the use of energy conservation
- The possible use of waste heat
- Adjustment to existing electrical installations
- Ease of maintenance along with availability of spare parts and consumable supplies
- Site accessibility

Despite the diversity of existing machines, a few general conclusions can be drawn about wind–diesel systems:

1. Assuming sufficient wind, it is technically possible to maintain an adequate power supply by wind turbines alone.
2. If no energy storage were installed, the system performance would be relatively poor, since oversized turbines would be required and a great deal of the wind turbine power might have to be dumped.
3. In order to achieve maximum performance, it is important to optimise the size of individual components along with any strategies for control and operation.

A schematic example of an AC wind–diesel system is shown in Figure 7.3. The wind turbine is a conventional stall-regulated turbine with an induction generator. The diesel engine drives a synchronous generator (sometimes called an alternator) through a controlled clutch. The system frequency is controlled by the dump load. If the output power of the diesel and wind turbine exceeds the demand on the local grid the dump load is gradually switched in and limits the rise in frequency. The voltage is controlled by the exciter of the synchronous generator.

At times of high wind speed and low load demand, the diesel engine is shut down and the clutch decouples the synchronous generator from the engine. The synchronous generator now remains connected to the electrical system to provide reactive power and hence voltage control. All the real power is then

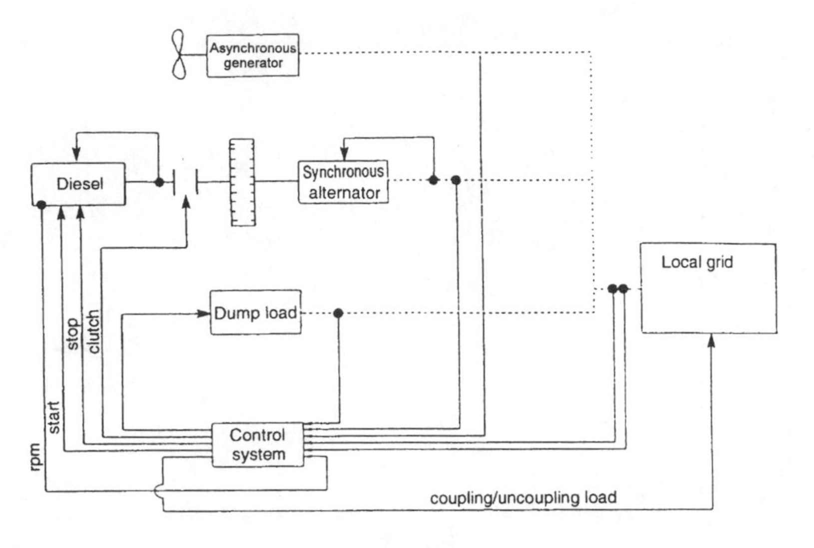

Figure 7.3. Schematic layout of a wind–diesel system

supplied by the wind turbine. When the consumer load increases, or the wind drops, the diesel engine is switched on, the clutch closed and the diesel generator starts to supply real power besides fulfilling its function of voltage control.

The system is composed of a 55 kW wind turbine, a 30 kW diesel generator set and a 75 kW dump load. Experiments with the system indicate that good power quality can be achieved both in the diesel plus wind-powered mode and in the wind-powered stand alone mode.

A flywheel storage system can be coupled to the system via a frequency converter system. The utilisation of the wind turbine can then be considerably improved, as the power fluctuations are stored in the flywheel.

7.1.2 Fuel Saving

The amount of fuel that can be saved by connecting a wind turbine to a diesel-powered grid system depends on the control strategy of the system as well as the local parameters of wind conditions and loads.

Figure 7.4 indicates the fuel savings for a wind–diesel system. It shows the fuel consumption as a function of the annual wind turbine output. The fuel consumption is presented as the percentage of consumption without a wind turbine, and the wind turbine output is given as a percentage of the annual energy demand.

7.1.3 Summary

This section has listed some of the considerations when designing wind–diesel systems for electricity generation. It discussed the desirability of

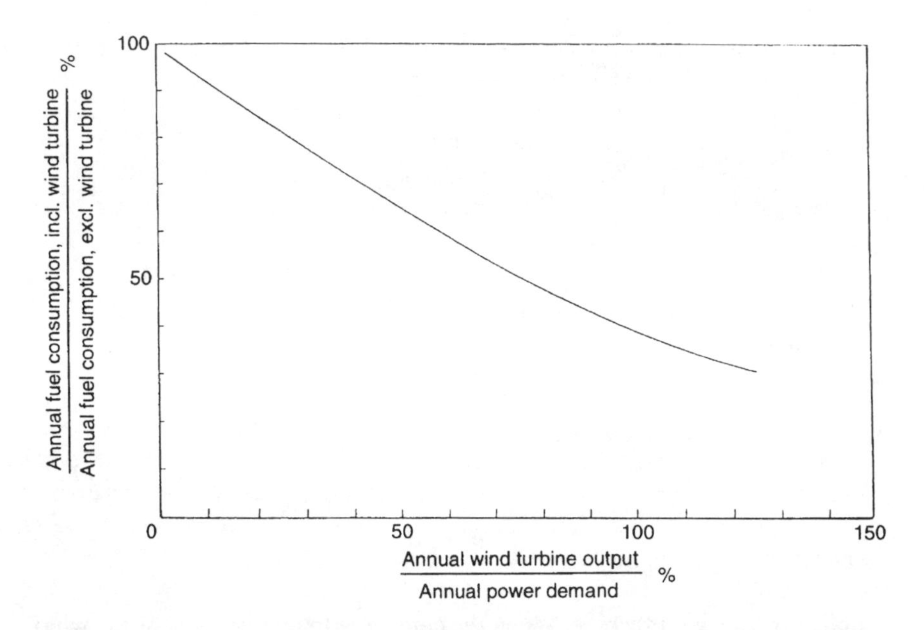

Figure 7.4. Fuel consumption for a wind–diesel system without storage compared to the consumption for a diesel-only system: 100% wind turbine availability is assumed (Reproduced by permission of the International Energy Agency).

incorporating some means of energy storage and the need to optimise individual components in order to achieve minimum system costs.

7.2 WIND TURBINES FOR WATER PUMPING

7.2.1 Introduction

Wind turbines have been used since ancient times to drive machines to pump water in many parts of the world. Early windpumps in Persia and China were of the vertical-axis type and it is thought the technology arrived in Europe from the eleventh century onwards. The traditional European wooden windmill with four sails or blades first appeared in the twelfth century and was widely established by the seventeenth century to pump water in the Low Countries. Technical developments were driven in the nineteenth century by the European colonisation of North America, where windpumps for cattle farms and water tanks for the transcontinental railways were in great demand. Until recently the multibladed windpump on a steel lattice tower was a common sight in the United States and in countries like Argentina and Australia; it is still the model for machines being manufactured today.

The use of windpumps declined dramatically from the 1920s due to the economic depression and the use of electric motors or petrol and diesel engines to drive water pumps. Soaring energy prices in the 1970s and growing interest in renewable energy sources led to their reconsideration, particularly in developing countries, although the take-up of the technology is still slow. In India, with full government support, there are now over 2000 modern windpumps used mainly for irrigation but also for village water supply and pumping seawater for salt production. It is estimated that throughout the world there are over one million windpumps in operation.

Poor quality drinking-water is the cause of more than half of human sickness in the developing world. The provision of pumped clean water is one of the best ways to improve health and increase the productive capacity of the population. Rural access to clean water is best achieved through pumping from underground water aquifers rather than using surface water sources, which are often polluted. Because of the relatively small quantities of water required, windpumping for village supply and livestock watering can be cost-effective given a good wind site. Irrigation pumping, however, requires large quantities of water at specific times of the year. For much of the year the pump may be idle or oversized and windpumping for irrigation may be more difficult to justify on economic grounds.

The main competitors to windpumps are diesel pump sets and solar pumps. Using a diesel engine costs less to buy and is easier to site, but requires frequent maintenance, an operator and the diesel fuel. Combining diesel with wind power as described in the previous section may bring benefits in some situations. The use of solar energy to power electric pumps has become a reliable and simple technology, and more details are given by Markvart (1994). Solar panels or photovoltaic modules convert the sun's radiation directly into electrical energy to drive the pump set. Solar pumps are expensive to buy initially but require little maintenance and no fuel. As with windpumping, the relative economics depend on the level of the resource, i.e. sunlight, at a particular site.

7.2.2 Design

The design of wind turbines for water pumping is relatively simple compared to electricity-generating turbines. The mechanical power at the rotor shaft is used directly to drive a pumping device. Turbines with high starting torque are suitable for pumping, and this requires high solidity rotors operating at low tip speed ratios of 2 or less. Most wind turbines for water pumping are of the horizontal-axis type and typically have rotors with 12–24 blades, diameters of 2–5 m and a hub height of 10–30 m. The rotor blades are usually made from curved sheet metal and need not be of a complex aerofoil section.

The usual pump head ranges from 10 to 150 m of water, depending on the size of the machine. Although centrifugal and screw-type pumps have been used in windpumping systems, particularly for low head, high volume

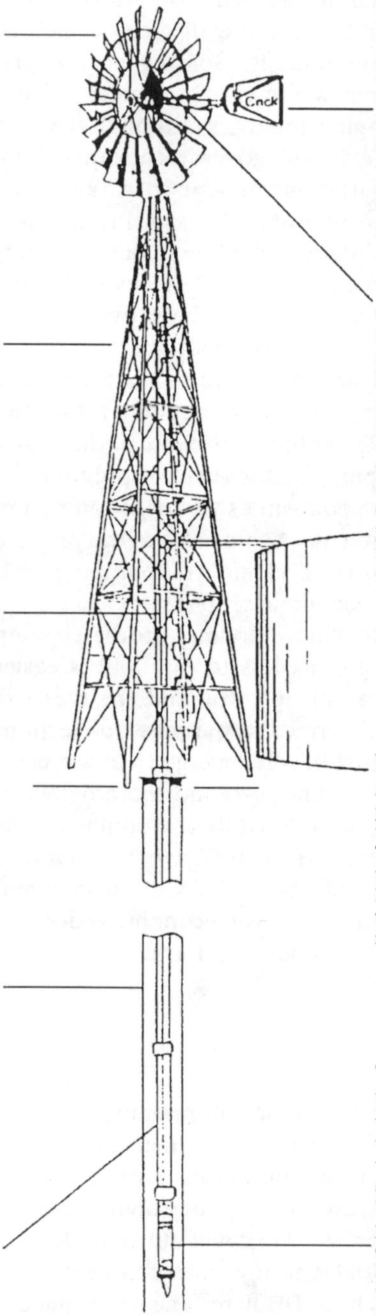

The rotor
This can vary widely in both size and design. Diameters range from less than 2 m up to 7 m. The number of blades can vary from about 6 to 24. In general a rotor with more blades runs slower but is able to pump with more force.

The tower
Normally metal (galvanised steel) with three or four legs. May be anything up to 15 m in height, but usually about 10 m. The bases of the legs are fixed, often by bolting to concrete foundations.

The pump rod
This transmits the motion from the transmission at the top of the tower to the pump at the bottom of the well. The motion of the pump rod is reciprocating (up and down) and the distance it travels (called the stroke) is typically about 30 cm, depending on the pump. Pump rods are usually made of steel.

The well
With a shallow water table the pump may be mounted on a hand-dug open well; if the level is deep a borehole must be drilled. The outer walls of the well will be lined to prevent infill but the liner should be slotted to allow water to enter the well.

The rising main
This is the pipe through which the water is pumped, and also encloses the pump rod.

The tail
Keeps the rotor pointing into the wind, like a weather vane. The whole top assembly pivots on the top of the tower, allowing the rotor to face in any direction. Most machines incorporate a mechanism into the tail which will turn the rotor out of the wind to prevent damage when it becomes too windy.

The transmission
Turns the rotation of the rotor into reciprocating motion (up and down) in the pump rod. Normal types use a gearbox or are direct drive. With direct drive the pump rod moves up and down once for each turn of the rotor. Using a gearbox allows the pump to be geared down so it does fewer pumping strokes for a given rotor speed, but with a larger output per stroke.

Wellhead components
At the wellhead the water is piped away to a storage tank through the discharge pipe. The pump rod usually runs through a seal at the wellhead, called a stuffing box, which prevents water escaping aroung the rod.

The pump
Normally submerged below water level. On the downward stroke the cylinder fills with water; on the upward stroke the water is lifted by the piston up the riser pipe. Pumps come in various cylinder bores and stroke lengths. The pump hangs on the rising main.

Figure 7.5. Anatomy of a windpump (Reproduced by permission from P. Fraenkel *et al.*, 1993, *Windpumps: A Guide for Development Workers*, Intermediate Technology Publications, London)

situations, piston pumps are much more common as they are more efficient and less expensive. Piston pumps should be used for heads of more than 10 m, and for heads greater than 50 m they can be over 90% efficient. A piston pump consists of a brass cylinder liner and a hollow piston with leather cup-seals (Figure 7.5). A disc valve in the piston allows water to pass upwards through it on the downstroke, but closes to lift the water above it on the upstroke. The brass cylinder is fitted inside a steel pipe, which in turn screws into the lower end of the rising main, a steel pipe extending from below the water surface to the top of the borehole. This arrangement allows the piston to be drawn up out of the rising main to change worn seals, usually every 1–2 years. At the bottom of the brass liner is the foot valve, normally closed unless the piston is on its upstroke and water is being drawn through the valve into the liner. The foot valve prevents the water in the rising main from draining down into the borehole when the windpump is not working.

To achieve the optimum efficiency of the combination, the performance of the turbine and the pump must be considered together. When starting a piston pump that is directly connected to the rotor shaft by a crank arrangement, a large force is needed to lift the pump rods and piston, to lift the weight of water sitting on the closed valve in the piston, and to overcome friction. The amount of water to be lifted is a column of water having the same diameter as the piston and the same height as the static head, which is the depth from the point of delivery to the water surface in the well. Once the rotor has moved through more than half a revolution from the bottom dead centre position of the crank, the torque required to move the piston decreases and there is sufficient momentum to smooth out the load as the rotor gathers speed. When the crank moves over the top dead centre, the weight of the pump rods plus water then acts to pull the rotor round, and the valve opens in the pump as the piston descends to admit more water.

A large starting torque is therefore required for a windpump but the mean torque that is necessary to maintain operation is only one-third of the peak torque. The minimum wind speed to maintain operation is only about 0.6 of the starting wind speed, as the torque is proportional to the square of the wind speed. The torque requirement of a well-designed piston pump is constant regardless of the speed of operation. The size of the pump determines the starting wind speed for a given wind turbine and pump head. The bigger the pump, the larger the starting torque requirement, hence the higher the wind speed to start the rotor. But piston pumps with larger diameters do have greater output because more water is delivered per stroke. Windpumps should be designed so they start in a wind speed that is approximately 0.75 of the local mean wind speed. This produces the best compromise between getting the windpump to run frequently enough and achieving good output in stronger winds.

The rotor needs to be kept aligned into the wind and has to be shut down when sudden increases in wind speed may cause overspeeding. Both tasks are carried out automatically by a large tail vane attached behind the rotor. The tail vane is hinged and is held in the normal position at right angles to the rotor by a tensioned spring. When the wind speed rises above the safe level the rotor yaws and increases the load on the tail vane; this overcomes the tension in the spring. It causes the rotor to swing round so the rotor disc lies parallel with the tail vane and therefore with the wind. When the wind speed drops, the spring is activated and the rotor turns back into the wind. Most windpumps are set so they automatically turn out of the wind, or *furl*, in the wind speed range 8 m/s to 12 m/s. There is also provision for manual furling if pumping is not required or the windpump has to be stopped for maintenance. A manual friction brake acting on a drum at the rear of the rotor shaft is a sign of good design, but it should only be applied when the rotor has furled and stopped rotating.

7.2.3 Transmission Systems

It is important to have a robust and reliable transmission system for carrying the forces generated by the rotor through to the pump; this is because it has to operate through many cycles in its lifetime. With a mechanical transmission system, rotors larger than about 4.5 m are directly linked to the pump crank, but smaller rotors which rotate faster use a stepdown gearbox with ratios of 2:1 to 4:1 as the pump speed should be limited to about 50 strokes per minute. Rotors with a geared transmission can thus run up to 200 rpm but directly coupled pump rotors are limited to 50 rpm. Limiting the pumping speed to 50 strokes per minute is necessary because cavitation and shock loads in the water supply can damage the system if the column of water in the rising main is accelerated too rapidly. At higher pumping speeds the pump rod may buckle on the downstroke if the mechanism tries to accelerate it rather than letting it fall under its own weight. In this situation, even if the pump rod does not buckle, there is increased wear on all the bearings due to the rapid reversal of the loads on the components.

In cases where the windpump has to be sited away from the water source, usually to improve wind energy capture, electrical, pneumatic and hydraulic transmission systems have been used between the wind rotor and the pump. A standalone wind turbine driving an electrical generator, which provides power for an electric motor driving a pump, is typically half as efficient as a good mechanical windpumping system. An AC generator with a rectifier and a battery can power either a DC pump motor via a regulator, or an AC pump motor via an inverter. The battery, which needs to be of the special deep discharge type, smooths the variation in output from the wind turbine and gives some storage capacity. A more recent system uses a high efficiency, permanent magnet AC generator to synchronise with an induction motor driving a centrifugal pump. This arrangement may be well suited for low lift, high volume pumping schemes where a mechanical windpump is limited by pump diameter and rotor size.

A few commercially available windpump packages use pneumatic transmission. There are obvious inefficiencies in a system which involves the rotor driving an air compressor, air transmission down a hosepipe then expansion through a slave pump actuator, pneumatic pump or an air motor driving a pump. The advantages are that the transmission line can be a low pressure pipe or hose, less expensive than electric cable; and the transmission is 'soft' so it can withstand shock loads between the wind rotor and the pump. An alternative to an actuated pump is the air-lift pump; it works on the principle that compressed air injected into the rising main, produces a froth of water. The froth rises up the pipe because the density of the mixture of air and water is less than the density of the water alone. However, lift pumps work best in deep boreholes where the water depth is more than the static head, i.e. the distance from the water surface to the delivery point. Hydraulic transmission systems where oil or water is used as the transmission fluid instead of air are even rarer than pneumatic systems.

7.2.4 Sizing a Windpump

The size of a pump driven by a wind turbine is a function of the pump head, the required water flow rate and the mean wind speed. This section outlines the steps that should be taken to obtain an estimate of the diameter of the wind turbine; this diameter determines the cost of the project and can be the basis of optimisation and economic assessment. It is assumed that the likely site has been surveyed and the probable position of the components relative to the borehole, river, buildings, etc., has been decided. The procedure follows Fraenkel *et al.* (1993), which should be consulted for its wealth of practical details.

The three main parameters that are needed are the total pumped head H (m), the pumped volume flow rate Q (m^3/s), and the expected mean wind speed V (m/s). The actual delivered power of the rotor must equal the required hydraulic power, so

$$C_P\left(\frac{1}{2}\rho A V^3\right) = \rho_w g H Q \tag{7.2}$$

where C_p is the coefficient of performance or efficiency of the rotor, ρ is the air density taken to be 1.2 kg/m^3, A is the rotor area in m^2, $\rho_w = 1000$ kg/m^3 is the density of water, and $g = 9.81$ m/s^2 is the acceleration due to gravity. Rearranging gives

$$A = \frac{1000 \times 9.81 HQ}{0.6 C_p V^3} \tag{7.3}$$

and the rotor diameter follows from $D = \sqrt{(4A/\pi)}$.

The required water flow rate, the pump head and the wind speed will vary throughout the year, so it is convenient to estimate the mean of each variable for every month. The daily demand for water is found from the likely per capita consumption either from field measurements or from internationally accepted minimum values. The head is the sum of the depth of the static water level below the surface, the maximum expected drawdown (see below), the height of the storage tank (if any) above the surface, and any dynamic head losses in the delivery pipes. The wind speed should be measured at hub height or the standard height of 10 m. The rotor area is then calculated for each set of monthly values. The month which needs the largest rotor area is called the *design month*. This represents the worst case, for if the system can meet the requirements for this month it can meet them for every other month. A value for C_p has to be chosen for these calculations and a convenient but realistic value would be 1/6 or 0.167.

If the rotor diameter indicated by the design month is greater than 8 m, normally the maximum for a windpump, then either the flow rate data should be checked, or consideration should be given to using more than one windpump or other types of pumping scheme.

An alternative way of sizing the windpump is to examine graphs of typical machines working under various conditions. An indication of the annual energy output (kWh) from typical sizes of windpump is shown in Figure 7.6, and the average water flow rate as function of wind turbine energy output and pump head is given in Figure 7.7. The energy density in the wind per unit area (W/A) in Figure 7.6 is related to the wind speed (V) by $W/A = 0.6V^3$.

Notice that both the water and the wind resource need to be quantified at the site where the windpump is to be erected. If hydrological data is not available, test boreholes may have to be drilled to determine the depth of the water table. As water is pumped from the borehole, the water level will drop

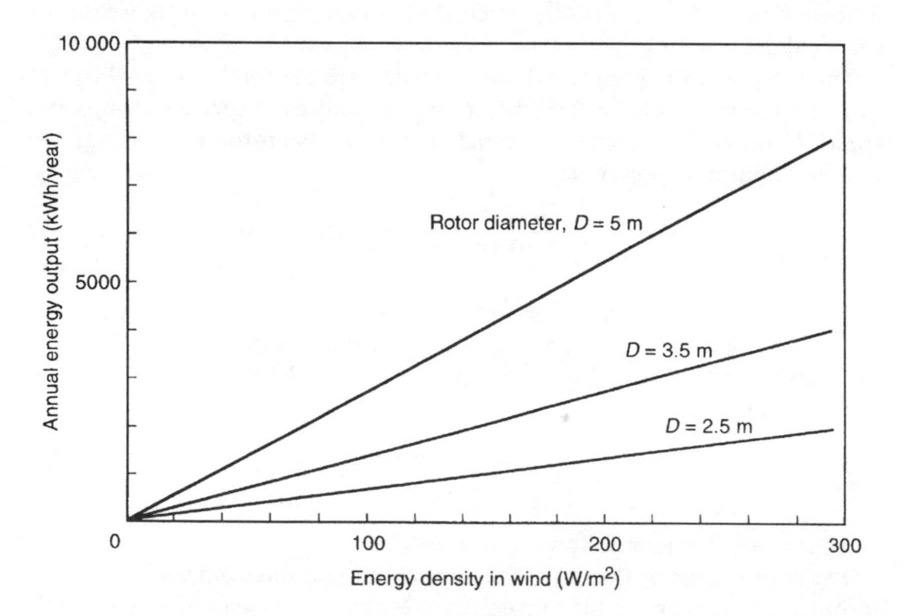

Figure 7.6. Indication of annual energy output for different sizes of water-pumping wind turbines as a function of the wind energy density

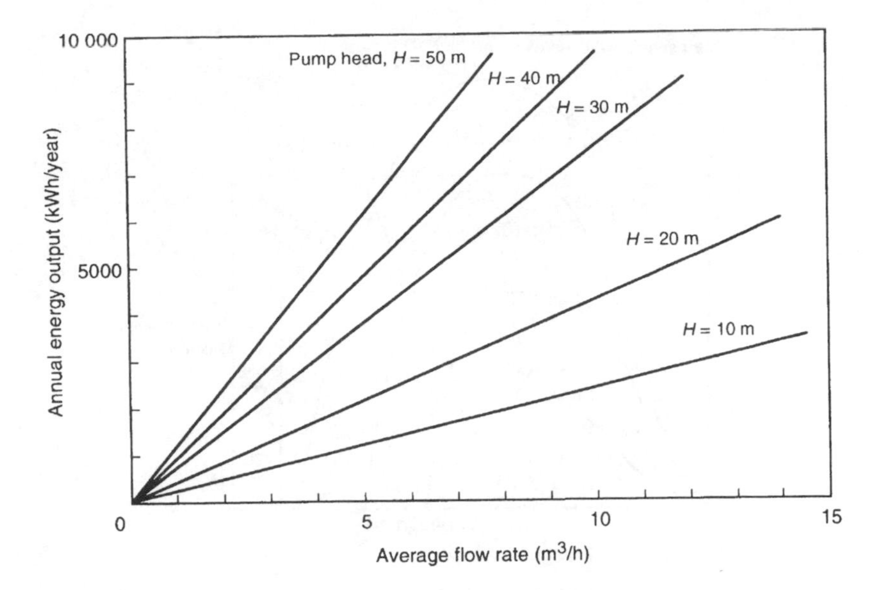

Figure 7.7. Correlation between annual energy output from the wind turbine and the water flow rate as a function of pump head.

below the surrounding water table, and this difference in levels, called the drawdown, has to be taken into account when estimating the total pump head. The windpump should be sized so that even in the windiest conditions the flow rate pumped out of the borehole is not greater than the inflow into the borehole from the water table. This is usually not a factor in irrigation schemes, where the water source may be from canals, rivers, land drains or shallow wells.

The daily water flow rate required for human consumption in a village or for cattle watering is easier to estimate than for irrigation purposes. The required flow rate will vary for different crops and types of soil and throughout the seasons. The amount of rainfall to be expected must also be taken into account. Further details of the complex calculations which may be required are given by Fraenkel *et al.* (1993)

The speed of rotation of the rotor of a windpump varies with the wind conditions and pumping requirement. Figure 7.8 shows how the load lines of a constant-torque piston pump intersect the power–speed and torque–speed operating curves of a windpump at various wind speeds to give the operating point. Variable-speed operation allows pumping over a wide range of wind speeds.

7.2.5 Summary

Some of the design parameters for a wind turbine driving a water pump were discussed with particular reference to developing countries, where the use of

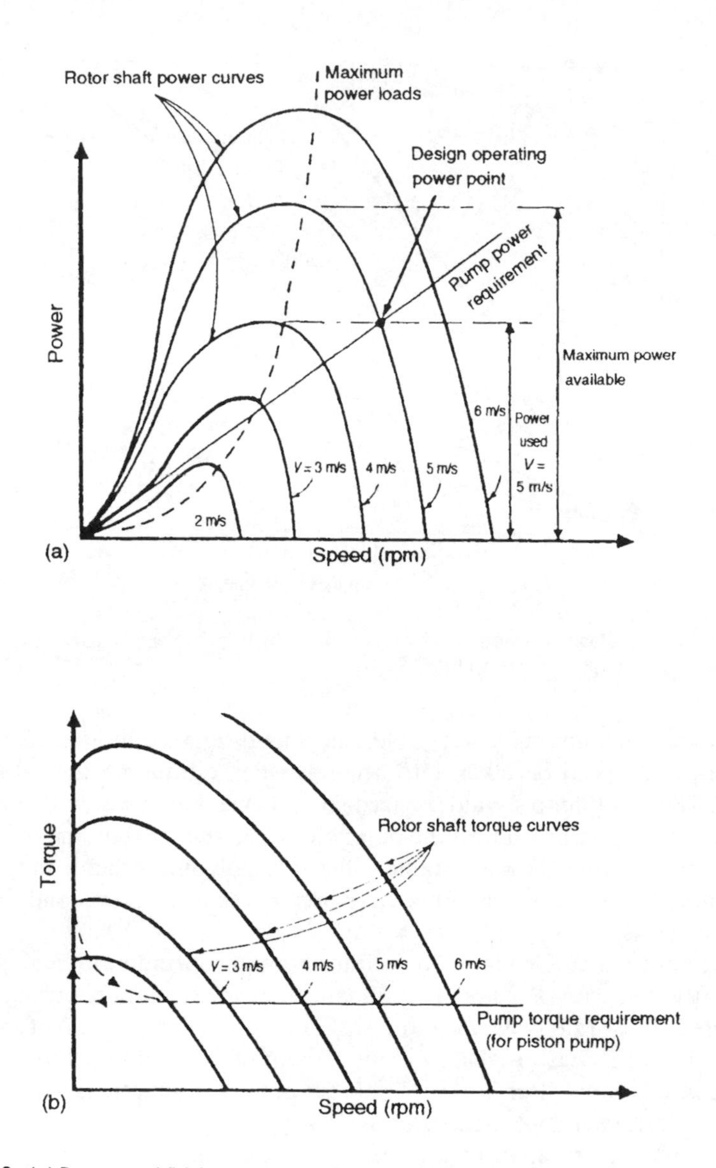

Figure 7.8. (a) Power and (b) torque curves for a windpump showing the effect of variable speed operation (Reproduced by permission from P. Fraenkel *et al.*, 1993, *Windpumps: A Guide for Development Workers*, Intermediate Technology Publications, London)

water pumps is growing. Emphasis was placed on the need to match the turbine's operating characteristics with the parameters of the pump. Piston pumps are commonly used and they require a large starting torque. To prevent the dangers of overspeeding with a piston pump in high wind speeds, the wind pump needs to be sized correctly and it requires a rotor-furling mechanism. An outline was given of the steps to follow when estimating the size of a windpump for a particular duty.

7.3 OFFSHORE WIND TURBINES

There is considerable interest in using wind turbines offshore and some European countries have set targets for the installation of offshore wind turbine capacity. Siting wind turbines offshore is attractive partly because of the higher wind speeds over the sea but mainly because of the reduced environmental impact. Two main restrictions on the siting of wind turbines on land are visual impact and noise. Both of these problems can be avoided by locating turbines some distance offshore.

The main disadvantage of offshore siting is the higher cost. Significantly higher costs will be incurred in offshore wind farms, particularly in terms of foundations, power collection cables and installation. As the turbines will need some additional protection from the harsh marine environment operation and maintenance are also likely to be more expensive. Various studies have indicated that the economics of offshore wind turbines improves with turbine size as the number of turbines reduces for a given energy output. Thus, the large wind turbines, of up to 2 MW rating, which are now becoming available should allow more economic offshore wind farms.

In spite of the rather poor economics of offshore wind energy at present, three wind farms were constructed up to 1996 and a several other installations are planned. The wind farms already constructed are at Vindeby and Tuno Knob, both off the coast of Denmark, and Lely in the Ijsselmeer inland sea in Holland. These are described in more detail in section 7.3.5. A number of wind farms are located on sea walls and harbour structures.

All three offshore wind farms already constructed are in rather shallow water; this is to minimise the foundation costs. In the future it may be necessary to locate wind turbines in deeper waters, and several innovative concepts have been considered, including the possibility of floating structures on which wind turbines are fixed. Floating structures create problems with mooring and power collection but they could be a way to harness wind resources in areas of deep water.

7.3.1 Offshore Considerations

Designing offshore wind turbines is similar to designing land-based equipment. The designer needs to determine wind turbine performance, structural response and optimum construction, but operation and maintenance schedules need to accommodate the hostile offshore environment. In addition to assessment of maximum wind speed, mean wind speed, turbulence intensity and wind shear, this requires determination of the structure of wind over the sea, land-to-sea wind speed recovery and prediction of hourly mean wind speeds at the site.

Wind speed data from a variety of sources need to be correlated using a suitable wind speed model, so the wind speed distribution can be determined for a particular site. There is also a need to consider array effects. Annual average wind speed in Europe for offshore locations is about 8.5–9 m/s. Technically, it should be possible to define wind speeds with an accuracy of 7 or 8% over the relevant areas of Europe.

Data can be obtained from a variety of sources, including lighthouses, oil platforms and islands, using a range of measuring techniques. Besides the variation of mean annual wind speed, an extra factor of importance is the temporal variation, the variation from year to year.

Wave data are required so that the response of the wind turbine structure to wind and wave loading can be determined. In some locations, breaking wave effects will need to be taken into account, depending on the sea depths at the wind turbine site. The effects of currents may also need to be considered. This is significant in the design of certain lighthouses.

Ice loading can be an important consideration in areas such as the Baltic; it should be taken into account.

Wave height distribution, like wind, is known to conform to a Weibull function, from which it is possible to determine maximum wave heights for design purposes. Consideration needs to be given to the possibility that waves will break when the water depth is less than about 1.3 times the wave height. Shifting sands might reduce depths locally.

In the UK, the total offshore wind resource without regard to economics is at least equal to the whole UK annual demand for electricity. Wind measurements have shown that the wind speed profile can be represented by a logarithmic law up to at least 80 m. The exponent was found to be 0.087 between 10 m and 80 m, lower than is commonly assumed. This should be compared with the onshore values, typically 0.28 for land with many trees and forests (Table 1.1). Wind speed durations conform to the standard Weibull function, the shape and scale factors are determined by detailed examination of wind data from a sufficient number of meteorological stations.

Extreme winds for design purposes are in the region of 60–70 m/s. The design gust speed is taken to be the wind speed expected to occur together with a fault on the turbine once on average every 100 years. Since the exposure period for fault conditions is bound to be less than for the normal condition, the maximum gust speed encountered will also be less, on average.

Palutikof *et al.* (1985) showed that the annual variation on land, and by extension over the sea, can be substantial. For example, in the period 1898–1954 at a notional site, a 60 m diameter wind turbine would have had an output of annual mean power falling continuously from about 480 kW to about 325 kW from 1917 to 1933. The annual power available would then have remained in the region of 300 kW for about another 20 years. Estimates of wind speed for the period 1917–1933 would have given substantially higher available power than for the next 20 years, and estimates of the economic viability of offshore wind power must take into account such variability from medium-term climatic changes.

The environmental impact of wind turbines located offshore must be assessed as carefully as for land-based installations. Aspects requiring particular attention offshore include impacts on birds, impacts on fish and sea mammals and interference with electromagnetic communication systems. In general there is unlikely to be much impact of noise or visual intrusion for turbines sited some kilometres offshore. It is also necessary to consider how the turbines and foundations will be decommissioned at the end of their useful lives.

7.3.2 Wave Data

Burton (1985), in a UK study, derived the highest value of the extreme wave height of 14 m and the value of the normal wave height for all locations was taken to be 6.5 m. For UK conditions, British Standard BS 6235 can be used as the basis for determination of the design waves. Taking into account the changes resulting from different seabed conditions, maximum and normal design waves can be calculated. The UK work is a comprehensive study of the problems of implementing an offshore installation, much of which is relevant today, except that it was concerned with large wind turbines of 3 MW output. The Vindeby wind farm described in section 7.3.5 is a recently commissioned installation off the coast of Denmark; it has many features in common with the UK study.

7.3.3 Design of an Offshore Wind Installation

To determine the economic prospects for an offshore wind energy power plant, several aspects need to be considered:

1. Definition of the scale and general layout of the offshore wind installation.
2. Analysis of structural concepts for a range of site characteristics, with preliminary design of support structures, fabrication and marine installations.
3. Estimation of capital and operational costs, taking into account the expected technical lifetime of the project.
4. Evaluation of wind regime, turbine size, turbine type, system efficiency and yearly energy production.
5. Assessment of the economic viability of the offshore wind installation.

Comparisons with alternative methods of generation show there must be significant cost reductions if offshore wind power is to become economically competitive.

Site studies need to include consideration of the seabed geology of each site as well as shipping, defence, fishing, environmental and communications constraints.

The development of design specifications needs to consider the following factors:

- Site description, including specification of wind data and relevant environmental conditions
- Operational requirements
- Strength requirements
- Structural design, construction and erection, maintenance procedures and safety considerations
- Instrumentation and data collection
- Inspection and testing

Studies are needed into the following areas:

1. Examination of installation problems and their solutions, including geotechnical studies, transportation, placing, electrical installation. Consideration of whether the installation should be made in sheltered or more exposed waters.
2. Determination of the natural frequencies and vibration amplitudes of the complete wind turbine under wind, wave and occasionally ice loading. Comparisons should be made with predictions.
3. Studies of corrosion, erosion (of the machine itself and of its foundation), and salt deposition effects.
4. Testing of facilities for remote operation and monitoring of the wind turbine installation.
5. Studies on methods of access and inspection, procedures for maintenance and repair, including the requirements for jetties.
6. Requirements for auxiliary supplies.
7. Studies of environmental effects of offshore wind installations.
8. Comparison between offshore and land-based wind turbine installations.

7.3.4 Generic Studies

At first sight, few areas appear specific to offshore wind installations, although an obvious topic is the behaviour of materials in the offshore environment, particularly their characteristics in salt sprays. But further consideration has identified many more areas that may need to be covered.

Although substantial information exists concerning meteorological and maritime matters, we need to improve our understanding of the wind structure over the sea, and the interaction between waves and the offshore wind structure. More work is needed on the theory and practice of wind turbine structural dynamics offshore. A number of generic requirements have now been identified, including

- Fatigue tests on materials in salt sprays
- Wave loading, including consideration of breaking waves
- Reliability, availability, maintenance and inspection
- Array efficiency
- Electrical engineering requirements

A structural dynamics analysis of an offshore wind installation would include the following elements:

1. Examination and assessment of existing structural dynamics models used for the design of shallow-water offshore structures in order to determine their suitability to analyse an offshore wind installation design.
2. Analysis of the complex physical forces acting on an offshore wind installation using adaptations of suitable existing models.
3. Evaluation of the resulting models with subscale systems tests.

7.3.5 Offshore Wind Farms

Compared with land-based equipment, offshore wind turbines have unfavourable economics; nevertheless, the decision was taken in 1989 to build a demonstration offshore wind farm at Vindeby, Denmark. It consists of eleven 450 kW stall-regulated wind turbines (three of which are shown in Figure 7.9) mounted on reinforced concrete caissons in water depths of 2–6 m. The wind farm is located 1.5–3.0 km offshore. It generates about 12 GWh of electricity per year, approximately 20% higher than would be generated by a similar installation onshore. The overall cost per kilowatt-hour is about 60% higher and the capital costs are 83% higher than for a comparable onshore installation, but it is estimated that later wind farms could be constructed at costs perhaps 20% lower. This would produce a substantial reduction in the electricity cost.

The turbines have rotors 35 m in diameter at a hub height of 35 m and are mounted on a steel tower. The turbine was of a similar design to those used onshore but with several extra features: an airtight tower and nacelle with a dehumidifier system; the finish of all outer surfaces was to the highest

Figure 7.9. Offshore wind farm at Vindeby, Denmark (Photograph courtesy of Bonus A.S.)

anticorrosion standards; heat exchangers were fitted for the gearbox and generator; and additional cranes were installed for maintenance. A 10 kV/690 V transformer was placed in the base of each tower. A 10 kV submarine cable, buried in the seafloor using trenches, transported the power back to the land.

The Tuno Knob wind farm is located off Jutland, the western part of Denmark. It consists of ten 500 kW three-bladed pitch-regulated turbines with 39 m diameter rotors set on 40.5 m steel towers. The wind farm is some 6 km offshore and is located in water of depth 3.1–4.7 m. The expected yearly electricity production is 15 200 MWh, giving a capacity factor of almost 35%.

The Lely wind farm is located in the Ijsselmeer, a large inland sea in Holland. There are four 500 kW two-bladed wind turbines with active stall control of the rotors. Using the active stall technique, the wind turbine blades are normally left at a fixed pitch angle and power is controlled by aerodynamic stall. However, the pitch angle of the blades is changed during braking and also to adjust the stall characteristic of the rotor.

All three offshore wind farms built to date have been for experimental or demonstration purposes. Electricity generated by offshore wind farms remains more expensive than electricity from onshore installations. However, the offshore wind energy resource is so large and of such potential benefit that ways need to be found to harness it in a cost-effective manner.

7.3.6 Research Requirements

Although several offshore wind farms have been built using essentially land-based turbine designs, considerable research work needs to be completed before cost-effective wind farms can be constructed.

Data Collection and Compilation

Extensive data exists on wind characteristics. However, as for any wind installation, it is essential to make specific site measurements before defining the requirements for a particular machine. Although structures can be designed using existing wave data, not enough is known about waves in shallow waters such as the southern North Sea, and more work is necessary.

Generic Studies

As well as the obvious studies of material behaviour in a marine environment, more work is needed on topics such as wave loading in shallow waters, structural dynamics, operational and maintenance requirements, array efficiency and electrical engineering.

Economics

At the moment, offshore wind installations are not an economic alternative to other forms of electricity generation, including onshore wind power. However, environmental considerations may favour 'clean' technologies by the introduction of incentives for their development, perhaps derived from a carbon tax on the use of fossil fuels. This could dramatically change the economic position of offshore wind turbines. It would also change the economic position of onshore wind power and other renewable energy sources, with which offshore wind would also have to compete.

7.3.7 Summary

The essential differences between the onshore and offshore environments were described, including the characteristics of the wind and the effects of these differences on wind turbine installations. The implications of siting wind turbines in various depths of water were also considered, along with the effects of wave and ice loading on structural design. This section also considered the environmental implications of offshore sites.

SUMMARY OF THE CHAPTER

Where there is no centralised generation of electricity the combination of a wind turbine with a diesel generator set can be a practical and economic solution to providing an electricity supply in remote areas. Good power quality can be maintained, but some system to damp out fluctuations in energy input and demand, is usually required, e.g. a flywheel or battery storage.

The use of windpumps has been common in many parts of the world over many centuries. In developing countries they are a very practical alternative to relying upon diesel fuel or centralised electricity generation for driving pumps to provide clean water for villages, cattle, and perhaps for irrigation. Windpump rotors are generally multibladed and have a low tip speed to give a high starting torque. Piston pumps are commonly used and their operating characteristics have to be carefully matched with those of the rotor. Although pneumatic and hydraulic means have been adopted for power transmission between the rotor and the pump, direct mechanical coupling to a crank mechanism remains the simplest and most efficient method.

Last to be discussed were the implications of siting wind turbines offshore. The most important difference between on- and offshore siting is that wind speeds are higher than on land, so the annual electricity output is increased. But the increased earnings from offshore machines are counterbalanced by increases in capital and installation costs of the wind turbines and the electrical transmission system to the shore. Maintenance and operational costs are also higher in an offshore environment. The cost of electricity from an offshore installation may be as much as 50% higher than from onshore wind turbines. Offshore siting is a way of counteracting the adverse public perceptions of onshore machines, but they do generate their own environmental considerations, including the need to investigate ecological effects on the seabed and surrounding areas, and the avoidance of interference to shipping, defence, communications and leisure activities.

BIBLIOGRAPHY AND REFERENCES

BS 6235, *Code of Practice for fixed offshore structures*, British Standards Institution 1982

Burton, A. L. (ed.), *Report on Offshore Wind Energy Wind Assessment Phase IIB, Vol. I, Executive Summary*. Taylor Woodrow, Southall, UK, February 1985. (Obtainable from ETSU, Harwell, UK.)

Caddett, *The World's First Offshore Wind Farm, Vindeby (Denmark)*, January 1994. (Obtainable from ETSU, Harwell, UK; CADDETT is an international agency for renewable energy.

Fraenkel, P. Barlow, R., Crick, F., Derrick, A. and Bokalders, V. *et al., Windpumps: A Guide for Development Workers*, Intermediate Technology Publications, London, 1993.

Hunter, R. and Elliott, G. *Wind diesel systems, a guide to the technology and its implementation*, Cambridge University Press, 1994.

Markvart, T., *Solar Electricity*, John Wiley & Sons, Chichester, UK, 1994.
Palutikov, P. J., Davies, T. D., and Kelly, P. M., *The Variability of the Wind Field over the British Isles – the Implications for Wind Power Production*. University of East Anglia, Norwich, UK, April 1985.

SELF-ASSESSMENT QUESTIONS

PART A. True or False?

1. It is not possible to maintain good power quality in an isolated system using wind–diesel plant.

2. The performance of a wind–diesel system will be poor if there is no energy storage.

3. A high starting torque is necessary for wind turbines used for water pumping.

4. Piston water pumps are suitable for any head up to 150 m.

5. Windpumps are designed to start at the local mean wind speed.

6. Windpumps with small rotors usually need stepdown gearing to transmit motion to the pump rod.

7. Windpumps with rotor diameters greater than 8 m are quite common.

8. The exponent in the logarithmic law for the wind speed profile over the sea is larger than for the profile over land.

9. Offshore wind power is currently competitive compared to other methods of power generation.

PART B

1. Use the graphical method to estimate the annual output in terms of cubic metres of water for a water pump turbine with a rotor diameter of 5 m. The pump head is 25 m and the power flux at hub height is 200 W/m^2.

2. What is the coefficient of performance and the wind speed for the wind turbine in Question1? Assume the water flow rate is 9 m^3 per hour.

3. What is the starting wind speed if at this duty in Question 2 the windpump is running at the local mean wind speed? Estimate the hourly water flow rate at the likely minimum wind speed for steady operation, assuming there is no change in efficiency.

Answers

Part A

1. False; 2. True; 3. True; 4. False; 5. False; 6. True; 7. False; 8. False; 9. False.

Part B

1. From Figures 7.5 and 7.6 the flow rate is about 9 m^3 per hour or 78 840 m^3 per year

2. From equation (7.2) $C_p = 9810HQ/P$ where P is the power in the wind and $P = 200A$ where $A = \pi \times 5 \times 5/4$. Hence $C_p = 0.156$ after converting Q to m^3/s.

3. Wind speed $V = (200/0.6)^{1/3} = 6.93$ m/s. Starting wind speed $= 0.75 \times 6.93 = 5.2$ m/s. Minimum wind speed for steady operation $= 0.6 \times 5.2 = 3.12$ m/s. Minimum flow rate $= 9 \times (0.75 \times 0.6)^3 = 0.82$ m^3 per hour.

8

Project Engineering

AIMS

This unit sets out the factors concerned with studies of the feasibility of wind energy for a range of applications, the main activities required for the successful implementation of wind energy projects, and the factors which will ensure the optimum technical and economic operation of wind power plants.

OBJECTIVES

When you have completed this unit you will be able to do three things:

1. Appreciate the various steps needed to evaluate the case for wind power in specific applications.

2. Understand how to plan the implementation of a wind energy project.

3. Develop an awareness of the, operation and maintenance routines of a wind power plant.

8.1 FEASIBILITY STUDIES

8.1.1 General

The decision whether to implement a wind energy project should be based on a feasibility study. The extent of the study depends on the size of the actual project. If the project deals, say, with a single water-pumping wind turbine, the study can be limited, whereas large wind farm projects require comprehensive studies.

The purpose of a feasibility study is to evaluate a project based on information about all aspects of the implementation and operation of the project. The data to be collected during the study can be divided into different groups:

Technical data

- Wind resources
- Electrical system and agreements
- Land availability
- Soil conditions
- Load pattern

Economic data

- Budget for implementation expenses, including capital costs
- Operation and maintenance budgets
- Present-value calculations
- Financing

Organisational data and information

- Organisation chart for implementation phase
- Organisation chart for operation and maintenance activities

The procedure for some of these activities is outlined in the following sections.

8.1.2 Technical Aspects

Wind resource assessment

As the very first step, it is possible by subjective observations to evaluate whether a site is windy or not. This can be done by considering such factors as whether the site is sheltered by the terrain, trees or buildings, or is favourably placed on a ridge, hill or coastline. The local population could be a valuable source of advice.

However, a reliable assessment of the wind energy resources requires that wind data are available for the actual area. Sources such as the *European Wind Atlas* or national databases should be consulted if data are not available for the site itself. This is then followed by consideration of two possibilities:

1. Setting up wind monitoring equipment at the site. Measurements of the wind should preferably be made at the proposed hub height, although measurements at 10 m above ground (meteorological standard) may be sufficient for small schemes. Both wind velocity and wind direction should be registered continuously as 10 minute or 1 hour average values.

2. Computing the wind condition based on data from a nearby existing wind meteorological station. The reliability of the data should be carefully evaluated and the influence of nearby obstacles and surroundings on the measurement equipment should be analysed. Long-term wind data

covering at least one year, preferably more, are required in order to make reliable wind resources estimates as there will be both seasonal and annual variation in the wind velocity. On its own, this second approach should only be considered in exceptional circumstances since it carries significant risk of providing misleading data.

For major wind farms the usual method of predicting the long-term performance of the project is known as *measure-correlate-predict*. It is a combination of the two approaches described above. One or more meteorological masts are installed at the site to *measure* mean wind speeds and direction as near to hub height as is practicable. This data is then *correlated* with the wind speeds measured simultaneously by the nearest recording station(s) of the national meteorological organisation. The correlation is usually carried out in twelve 30° sectors to take account of directional effects between the meteorological station and the wind farm site. In this way twelve 'speed-up factors' are established to relate the mean wind speeds at the site with those of the meteorological stations. Finally the long-term wind speed at the site is *predicted* from the long-term records of the meteorological station.

The advantage of this method is that it allows the very long-term records of the meteorological station – sometimes up to 40 years of data may be available – to be used, so it limits the risk of choosing a particularly windy year in which to make site measurements. It also provides an experimental method of relating the long-term data to a site which may be many miles from the meteorological station.

A final step of wind farm energy assessment is to predict the wind speed at the locations of all turbines, locations which may be some distance from the meteorological masts. Array effects and losses must also be taken into account. Several computer codes are now available to do this, taking account of local topography and the effect of other wind turbines. These work effectively within the region of a wind farm but may fail to give accurate results over long distances and in complex terrain.

As the energy of the wind is proportional to the cube of the wind speed, it is of great importance that the wind speed at a certain site is well known. The commercial risks of incorrect resource assessment are so large that considerable time and effort should be spent on this aspect of the feasibility study.

Electrical grid system

If a project concerns turbines for connection to an electricity network, the utility will require evaluation of the power quality and design of the wind turbines, particularly with regard to control and protection of the plant. This will usually be a legal requirement. Some of the technical aspects of the connection to the public supply network were discussed in Chapter 6. However, the legal and commercial agreements are also particularly important during the initial stages of the project.

The connection of any privately owned generator to the public electricity supply network is likely to require a number of formal agreements or contracts. The most obvious requirement is to define who is to buy the energy (kWh) generated by the turbines and at what price. This is usually defined in the *power purchase agreement* and is likely to be with the host electricity utility or a government agency. Alternatively, in a deregulated electricity supply industry the energy may be 'wheeled' to other customers. One difficulty with wind generation is that all the main costs are incurred by the developer during the construction of the project then revenue is received over a number of years. Thus it is highly desirable to have a guaranteed market for the electricity for the life of the project, perhaps 15–20 years, in order to reduce the risk of the developer and those who have invested in the project. A power purchase contract of this length may be difficult to obtain. A second difficulty occurs if the wind power is 'wheeled' through thc network to a particular consumer. As the power output from a wind farm is intermittent, any consumer using wind-generated electricity will also have to purchase electricity from other sources during times of low wind speed. This naturally reduces the value of the wind-generated power as the alternative supply is likely to come at a premium.

In addition to exporting real power (kW) a wind farm is also likely to import reactive power (kVAr). A small amount of real power will be required when wind speeds are low, in order to supply the wind turbine controllers, hydraulic pumps, cabinet heaters, etc. An agreement needs to be made with the local utility for the supply of this real and reactive power. This may be of the usual form for the supply of electricity to a small industrial unit. However, care is required as a typical 10 MW wind farm may absorb 3 MVAr at maximum output and only 300 kW during low winds. Many tariffs designed for industrial supplies may not be suitable for such a load.

The wind farm also needs to be connected to the public electricity network, which will probably require the construction of new circuits. This is usually defined in the *connection agreement*. When charging for connections the utility may adopt two basic philosophies. In *shallow charging* the generator has to pay only the costs of the circuit to the nearest suitable point of the distribution system. The costs of any additional system reinforcement are borne by the utility then recovered over time by levying *use-of-system charges* on the generator. The alternative is the so-called *deep charging* approach, where the generator pays for all work required to allow the wind farm to be connected; sometimes the modifications to the electrical network are many miles from the wind farm and in deep charging they are still paid for by the generator.

The deep charging approach is more common on distribution systems but, is not entirely satisfactory. A common complaint is that a wind farm developer may pay for system reinforcement that benefits either the host utility or another generator. Work to reinforce the utility system usually has to be undertaken by the staff of the utility, so it may be difficult to confirm that the costs are reasonable. At present most connection charges tend to be worked out on a case-by-case basis and a degree of engineering judgement is used to assess who will benefit from the equipment installed. However, as generation plant embedded in distribution systems becomes more common, any charges for extensions to the distribution network will be made on a more formal basis.

Land availability

An important condition for implementation of a wind energy project is that land is available and the site is accessible. The directly required area of land depends on the size of the turbine. For example, the required area for the foundations of 200 kW grid-connected turbines is in the order of $100m^2$. In addition, approximately $200–300m^2$ are required for the control building and service area. Access to the site is an important issue, particularly with the large cranes required for modern wind turbines, and it may be necessary to arrange for improvements to the public roads as well as the construction of access tracks on the site itself. There is also a temporary requirement for a lay-down area of land around the base of each turbine. This is used for standing the cranes and laying down the blades towers and nacelles before they are erected; it can be returned to agricultural use once the installation is complete.

The electrical circuits connecting the wind farm to the public electricity supply network will also require land and access rights. Obtaining these rights for long overhead-line circuits may be a major constraint on the development of a project.

Ground conditions

The cost of the foundations depends on the ground conditions, so ground and soil surveys must be carried out at an early stage of the study. In general, simple slab foundations are used for the turbines and the size of the slab may be increased for poor ground conditions. Some sites have a surface layer of peat and this needs to be removed in the foundation and lay-down areas. The cost of the site access tracks will also be determined by the ground conditions and on some sites it has been possible to install tracks based on geotextile fabric so they 'float' on top of the surface soil or peat. Although rocky ground allows the use of small-diameter turbine foundations, it may increase the costs of trenching for the power collection cables.

Load

The output of a wind farm is usually sold under some form of power purchase agreement. However, for some small schemes, e.g. on some islands, this may not be the case. The load pattern and the costs of the electricity generated by the conventional generators must then be known, in order to estimate the value of the power produced by the wind power plant. The power produced during base load periods is less valuable than power produced during times of peak demand; this is because it has to compete with low cost conventional generation.

For turbines not connected to a grid system, the maximum benefit of the output is obtained if the load pattern is identical with the output variation. This applies to windpumps as well as electricity generators.

8.1.3 Economic Issues

The costs can be divided into the following parts:

- *Wind turbine equipment*: includes the capital cost of the wind turbines, control system, transportation, assembly and erection
- *Foundation work*: includes excavation and casting
- *Other civil work*: includes land, roads and buildings
- *Electrical work*: includes connection to the power system, grid lines and transformers

The costs to be included in a typical *operational and maintenance (O&M) budget* include

- Planned maintenance
- Unplanned maintenance
- Spare parts
- Insurance
- Administration
- Contingencies or unpredictable expenses.

Provision must be made for staff costs, either indirectly through contract operation or maintenance, or by direct employment.

The project will normally be evaluated using the net present value (NPV) technique described in Chapter 4. A discount rate will be chosen by the investors in the project to reflect their view on the value of their money. This discount rate may vary widely from a low value (5–8%) for publicly owned utilities up to (15–20%) for private investors. The higher the discount rate, the more difficult it is to develop a successful project.

Finally it becomes necessary to arrange finance for the project. National utilities and other very large organisations are likely to finance small projects such as wind farms from their own funds. Independent developers will need to arrange loans from banks and other lending institutions. One important issue is to decide whether the loan will be secured only on the project (known as non-recourse or project finance) or whether additional guarantees based on other assets are required. If project finance is used, the lenders will be very careful to ensure the project is well planned and the feasibility study is sound.

8.1.4 Organisational Aspects

It is common sense to establish a competent organisation for successful implementation and operation of a wind energy project. It is normal practice for funding agencies, finance houses and insurers to ask for evidence of the technical and organisational competence of those initiating the project. This usually includes a requirement that suitably qualified technical staff are available for operation of the plant to ensure maximum output and lifetime. The management of the project normally has to conform to appropriate legal requirements.

8.2 PROJECT IMPLEMENTATION

8.2.1 General

When the decision on implementation of wind power plant is taken by project planners, a number of activities involving various technical disciplines have to be initiated. The main engineering tasks from the developers viewpoint are to prepare a technical specification and to coordinate all the inputs to the project. For small projects with only one turbine the work is relatively simple, whereas wind farm projects with several wind turbines require considerable coordination and planning.

There are two distinct approaches to project development. The first is for the developer to approach an organisation that would implement the entire project under a single contract. This is commonly known as a turnkey contract. The second approach is for the developer to engineer the project itself . A combination of the two is possible. If the entire project is not to be turnkey, the developer must make provision for its own engineering and consultancy costs.

Two types of specification may be written. A *functional specification* seeks only to define the functions required of the equipment, not to constrain the supplier in the way these functions are achieved. As the details of the equipment are not specified it is very important to include clearly defined performance criteria and full details of how they will be tested.

The other type of specification is known as a *detailed specification*. Here all the details of the equipment are specified by the purchaser who then starts to accept the risk involved in the design of the equipment. Detailed specifications are useful when the purchaser has very specific requirements and the technical knowledge to specify them accurately. A disadvantage of detailed specifications is that suppliers may not be able to offer their standard products. Table 8.1 shows a highly simplified example to illustrate the two different approaches. The table illustrates a number of important points. For most applications it does not matter whether a turbine is pitch or stall regulated or the material used for the blades. Therefore, these decisions are better left to the manufacturer. Similarly, the speed of the generator is not

Table 8.1. Simple comparison of detailed and functional specifications

Functional specification	Detailed specification
Turbine shall produce 1 000 000 kWh/year at the site; site mean annual wind speed is 8 m/s	Turbine shall be 35 m in diameter, pitch regulated with glass reinforced polyester blades
The wind farm shall be suitable for connection to a 50 Hz, 33 kV network Testing to confirm wind turbine power performance will be undertaken by the contractor in accordance with IEA standard procedures	The generator shall be of the induction type operating at 1000 rpm The tower shall be 35 m to the yaw bearing, of steel construction and with an internal ladder
Interfaces with the foundation, electrical and other systems are clearly described and marked on the drawings	Interfaces with the foundation, electrical and other systems are clearly described and marked on the drawings

important as long as the frequency of the electrical system (50 or 60 Hz) to which the turbine is to be connected is clearly stated. What is of great interest to the purchaser is the performance of the turbine at the proposed site and how this performance may be verified by testing. Thus it would appear sensible to specify a wind turbine using a functional *specification*. However, if a particular design feature is required, perhaps for a wind–diesel system, a detailed specification might be appropriate. Whichever type is chosen, it is extremely important to specify the interfaces very clearly and any environmental or health and safety requirements that have to be met.

8.2.2 Project Phases

The project management activities are divided into phases, as illustrated in Table 8.2.

Table 8.2. Overall project phases and activities

Phases	Activities
Tender specifications	Specifications must be made on the basis of data for the actual site; it is convenient to divide the specifications into four technical areas: • Wind turbines • Foundations for wind turbines • Civil works (buildings, roads) • Electrical work
Contract negotiations	Negotiations with contractors regarding technical, economic and legal conditions for the contracts

Phases	Activities
Manufacturing and construction	Manufacturing of turbines and construction work at the site (Figure 8.1)
Installation	Erection and assembly of turbines; electrical installation of high and low voltage equipment
Tests	Before final takeover, a number of tests are made in order to check that the equipment fulfils the specifications; the tests include verification of performance and inspection of the installation
Commissioning	After fulfilment of the conditions for commissioning, the system is handed over to the end user

After commissioning it is recommended to follow up with a monitoring phase in order to check the performance of the plant during an agreed guarantee period in which suppliers are held responsible for defect liability.

After expiry of the guarantee period, satisfactory performance of the wind power plant during the lifetime of the plant may be ensured by appropriate standards of operation and maintenance. This can be done within the project

Figure 8.1. Work in progress on the foundations of a wind turbine at Delabole, Cornwall, UK

Figure 8.2. Tower assembly. Reproduced by permission of National Wind Power

organisation or through service contracts with the wind turbine manufacturer or other competent bodies.

8.3 OPERATION

8.3.1 Training

A wind energy project will not be successful unless the operation and maintenance (O&M) staff are adequately trained.

The extent of the training depends on the type of wind power plant as well as the experience of the staff to be allocated to the project.

In general training at two levels is relevant:

1. Theoretical training for professional staff with managerial functions in the organisation.
2. Practical training for skilled technicians who are going to have responsibility for the daily O&M of the plant.

The existence of courses leading to qualifications recognised by the wind turbine industry would be an advantage and would reduce the need for training to be carried out by the project organisation itself.

8.3.2 O&M Procedures

O&M procedures in order to secure effective operation and maintenance of the plant will need to be based on procedures specified by the wind turbine supplier and any other suppliers. Provision of the necessary information will be a condition of contract with the suppliers. Guarantees by suppliers will be conditional on evidence supplied by the project organisation that the recommended O&M procedures have been followed.

Figure 8.3. Rotor being lifted into position. Reproduced by permission of National Wind Power

Summary of the chapter

This chapter has discussed various important aspects concerning the engineering of wind energy projects. All projects need to start with a feasibility study and a very careful assessment of the wind resource at the site. A number of agreements with the landowners and the local power utility then need to be arranged. The developer must then decide whether he or she wishes to purchase a complete 'turnkey' wind farm or to let a series of contracts that need to be managed. In either case it will be necessary to write specifications that describe the work required from the contractors. The various stages of a project were then explained, along with the importance of an effective operation and maintenance organisation.

Bibliography and References

Troen I and Petersen El., *European Wind Atlas*, Riso National Laboratory, Roskilde, Denmark, ISBN 87–550–1482–8.

SELF-ASSESSMENT QUESTIONS

PART A. True or False?

1. A wind turbine project must always be preceded by a feasibility study.

2. It is not important to use detailed data on the wind resource when planning a wind turbine project.

3, Power quality is not an important consideration when wind turbines are connected to an electricity grid.

4. Each 200 kW wind turbine requires 1 km^2 of land.

5. Matching the load pattern is not a benefit if wind turbines are not connected to an electricity grid.

6. Because the wind is free, the cost of wind power is low compared with many other power station investments.

7. Because wind turbine plant is relatively simple, operation and maintenance by qualified technical staff is not necessary.

8. Comprehensive planning is required for larger wind turbine projects.

9. It is better to overestimate average wind speeds when planning a wind power project.

PART B

1. Why is it particularly important to obtain adequate data on the wind regime for a specific wind turbine site, particularly with regard to the probability of lower wind speeds?

2. What is the first activity in the implementation phase of a new wind turbine project?

3. Name the phases in a typical wind project up to the point when the end user takes over?

4. What two main activities must be carried out either directly or indirectly by the end user?

Answers

Part A

1. True 2. False 3. False 4. False 5. False 6. False 7. False 8. True 9. False.

Part B

1. Because the energy available in the wind is proportional to the cube of the wind speed, it is crucial to estimate wind speeds with sufficient accuracy. Overestimation would mean the wind turbine producing less energy than expected with an adverse effect upon the economics. Underestimation would cause the turbine to be undersized, so the potential earning capacity of the site would be reduced.

2. The first activity in a project is to produce a technical specification so that a potential supplier can understand the project requirements.

3. The six phases in a wind turbine project are listed in Table 8.2.

4. The end user must ensure that operational and maintenance activities are satisfactorily carried out.

9

National and International Wind Energy Programmes

AIMS

This unit outlines some of the national and international programmes which are in progress to harness the great potential of the wind resource.

OBJECTIVES

At the end of this unit you will appreciate the scale and nature of some of the wind energy programmes being implemented in various countries and regions of the world.

9.1 FORMS OF SUPPORT FOR WIND ENERGY

The utilisation of wind energy as a source of electrical power is, in many situations, apparently more expensive than using conventional fossil fuel plant if conventional financial appraisal methods are applied. The full external costs of fossil fuel in terms of acid rain and climate change are not generally taken into consideration when comparing generation options, and the problem of long-term depletion of fossil fuel reserves has too extended a timescale to be addressed effectively by conventional financial appraisal techniques. However, many governments and international agencies have recognised that the development of wind energy and other new renewable energy resources is so desirable that mechanisms must be found to encourage it.

In general, two main approaches have been adopted to support the development of wind energy: direct support for development of wind turbine technology and support for the market for wind-generated electricity. Early initiatives were mainly to support technology development, but as reliable wind turbines have become widely available, market support schemes have proven very effective. Support for the development of the technology has taken several forms including national and international research and development programmes, demonstration programmes to show commercial operation of a new design or concept and the establishment of suitable standards and codes of practice for design. Market support can be a the long-term guarantee of a fixed premium for wind-generated electricity, direct support for the purchase of wind turbines or tax relief on wind plant investments. Other administrative arrangements have been important in some countries, e.g. appropriate government advice and guidelines for the local planning authorities, suitable commercial arrangements to allow the wind-generated electricity to be transported through the public electricity network and appropriate legal frameworks. These support measures have varied widely from country to country and change with government priorities. However, they continue to have a major impact on the development of wind energy.

By mid–1996 there were some 5500 MW of grid-connected wind turbines in operation throughout the world, demonstrating the effectiveness of these support measures. The cost of wind-generated electricity continues to fall as turbine sizes increase and the technology develops. Over time it is likely the cost of fossil fuel generated electricity will rise as the full external costs are borne by the generator and fossil fuel prices rise. Once the costs of wind and fossil fuel generated power converge, then there will be no further requirement for government support measures and wind energy will take its place among the other technologies as a competitive form of electricity generation.

9.2 THE INTERNATIONAL ENERGY AGENCY

Worldwide collaboration on wind energy takes place through the International Energy Agency (IEA). Although it is the energy forum for the 23 industrialised countries of the Organisation for Economic Cooperation and Development (OECD), other countries are invited to participate in some of its activities.

The IEA programme on research and development of wind turbine systems began in 1977 and by 1993 had 14 contracting parties from 12 countries. The contracting parties are either governments or appropriate research and development institutions designated by governments. Individual projects or tasks are controlled by an executive committee on which each contracting party is represented. Recent and ongoing research, for example on turbine aerodynamics and component fatigue, is presented and discussed at yearly symposia. Meetings with experts are also organised to help communicate research results and to discuss programme plans.

A study of wind–diesel systems culminated in the publication of a handbook about their siting and implementation (Hunter and Elliot, 1994). The effect of wake interactions from turbines in wind farms has been the subject of another study. Many years of research indicate the array energy losses and turbulence effects in wind farms are significant but less than was first thought.

Ongoing tasks include the experimental study of a 16 m diameter two-bladed wind turbine on a test site in Germany. A large database of measurements has been built up since 1988 on the operational behaviour of different hub designs, stall control and the effects of rain and ice. Verification of a numerical simulation model for the whole turbine has also been carried out. Another important task is the issuing of reports on recommended practices for wind turbine testing and evaluation; these reports form the basis of international standards for wind energy technology.

Another task which has evolved since 1977 is concerned with the development of large-scale wind energy systems. Originally concentrating on the development of wind turbines having outputs of greater than 1 MW, the project now includes an examination of the issues that arise from operating large wind farms. Contracting parties come from Canada, Denmark, Germany, Italy, Japan, Norway, the Netherlands, Spain, Sweden, the United Kingdom and the United States.

9.3 THE EUROPEAN UNION

The European Union (EU), previously known as the European Community, has funded large programmes of research, development and demonstration of wind energy technologies. The Commission for the European Communities through its Directorate-General XII for Science, Research and Development has promoted joint opportunities for unconventional and long-term energy options (JOULE). The second programme of JOULE finished in 1994; it included work on wind energy and brought together universities, research centres, industry and consultants to work on specific projects under a self-selected system of consortia. The projects were agreed by recommendation of independent experts and were funded half by the Commission and half by the participating bodies. Today the participants are not confined to EU countries; countries from the European Free Trade Area (EFTA) and central and eastern Europe are also included.

The general objectives of the wind energy component of the JOULE programme are to develop advanced large wind turbines, to support research on unresolved problems of existing wind turbines, and to promote wind energy technology that will enable the use of unconventional sites such as those having complex terrain or lower wind speeds.

The first phase of the programme on large wind turbines was concerned with experimental machines and the results were published in a book by Hau *et al.* (1993). The current second phase to be completed in 1996 is oriented towards producing large turbines for the market and is taking the majority of the Commission's funding for the wind energy sector. Objectives include improving the cost-effectiveness by reducing the weight of the rotor and nacelle, and improving the ease of transport and erection. The advances in the technology are shown by the machines now being developed by the seven consortia which have specific weights that are half those investigated in the first phase. By mid-1996 there were four prototype commercial wind turbines rated at approximately 1.5 MW with rotor diameters of approximately 60 m in operation. European Union support has contributed to the successful development of most of these very large turbines.

Research on unresolved issues, known as generic research, includes new materials and fatigue properties of blades, gearless generators, unsteady stalled flow on blades, noise reduction and integrated design methods. Other projects bring together the electricity utilities and research institutions to exchange experiences and promote technology transfer of renewable sources of energy. For wind power this includes agreeing common specifications for the supply of wind turbines, overcoming problems relating to large penetration of wind power in utility networks, and obtaining planning permission for new wind energy projects. Future research is likely to be on rotor aerodynamics, development of wind turbine standards, and wind measurement and modelling in complex terrain and offshore environments. Developing a good understanding of wind behaviour on complex terrain becomes more urgent as the more benign onshore sites are exploited. Planning, environmental issues and public acceptability will also become more important for collaborative research.

Another arm of the EU which supports wind energy development is the THERMIE programme. Administered by Directorate-General XVII of the Commission, it deals with energy technology: strategy, dissemination and evaluation. THERMIE is concerned with the development and promotion of new energy technologies in order to improve security of energy supplies and to preserve the environment. The aim is to achieve commercial exploitation of the technology and funding is provided for innovative projects in a new market or when there is a possibility of a significant technological breakthrough. The dissemination of information is seen to be an important part of the programme and is carried out through technical seminars, technical fairs and appropriate databases. THERMIE has funded the creation of demonstration wind turbines and wind farms in the EU and has carried out market studies of the wind potential in island regions, the Baltic States and eastern Europe.

9.4 DENMARK

The country with the biggest proportion of its electricity produced by wind turbines is Denmark. Following the sudden increase in oil prices in the 1970s, the Danish government began to promote wind energy as a means to become less reliant on imported oil. Installation subsidies helped to increase the number of privately or cooperatively owned wind turbines, and they grew so substantially that by 1993 there were 3400 grid-connected machines with a total capacity of 450 MW. These were generating 800 GWh/year or 2.75% of the total electricity consumption. By 1996 the total capacity of wind turbines in Denmark had increased to some 700 MW. Although single machines still predominate, the number of wind farms began to increase from 1985 as the electricity utilities began to develop them with government encouragement. As the payback rate for wind energy has been so good, the level of public subsidy has been greatly reduced.

In the 1980s a typical Danish wind turbine would have a three bladed 16 m diameter rotor, stall regulated on a 20 m steel lattice or tubular tower and driving a 55 kW induction generator. By 1992 new machines had a capacity of 400–500 kW on 35 m towers. The government target is to have 1000 MW of wind turbines installed by the year 2000, producing about 5% of the country's electricity.

The development of large wind turbines is being actively pursued by the Danish government and the electricity utilities with financial support from the European Union. Two 40 m diameter, 630 kW turbines were commissioned in 1979 at Nibe in northern Jutland. The mean wind speed at this site is 7.1 m/s at the hub height of 45 m. One machine is stall controlled and the other is pitch controlled. The original fibreglass blades suffered failures and have been replaced by laminated wooden blades which are still under test.

A wind farm consisting of five machines similar to the Nibe prototype with pitch control was commissioned in 1987 at Masnedø, a small island off southern Sjaelland. Two of the rotors have been replaced with wooden blades and the other three have been fitted with new fibreglass blades. The machines have operated satisfactorily apart from gearbox failures.

The largest Danish wind turbine was commissioned in 1988 at Tjæreborg on the west coast of Jutland. This machine has a rotor of diameter 60 m and is rated at 2 MW. The rotor has three blades, the generator is of the induction type and the tower is concrete. This machine also suffered a gearbox failure one year after commissioning, but after repairs it has been operating satisfactorily.

Another large machine was commissioned in 1993 at Avedøre some 10 km south of Copenhagen. The rotor is 50 m in diameter and the machine is rated at 1 MW. The blades have been designed for both stall and pitch control, and measurements are being carried out to compare the two control systems.

The world's first offshore wind farm was commissioned in 1991 at Vindeby, northwest of Lolland in the Baltic Sea. The water depth varies from 2 to 6 m and the distance from the shore ranges from 1.2 to 2.4 km. There are 11 turbines in two rows, with each rated at 450 kW. The rotor diameter is 35 m and the hub height is 37.5 m. The availability is over 96% but the delivered cost of electricity is about 50% higher than for an equivalent onshore wind farm.

An important advance made in Denmark is an approval and certification system to improve the quality of Danish wind turbines. All wind turbines installed after July 1992 must meet requirements for documentation of all design criteria such as loading, fatigue evaluation, safety levels and power curves; proper documentation is also needed for quality procedures during manufacturing, transportation, installation and subsequent servicing.

Danish wind turbine manufacturers seeking export credit guarantees also have to obtain this certification. This is another indication of the serious commitment to wind power that exists in Denmark.

9.5 INDIA

Due to the large national power deficit, the government of India has included wind energy in national power plans and a number of studies have been carried out. The country had been regarded as a region with low wind speeds until a programme of wind mapping and monitoring initiated by the government identified several sites with wind speeds greater than 6 m/s. Some of the sites due to complex terrain effects experience wind speeds greater than 8.5 m/s. The potential areas are the coastal strips, inland gaps and ridges in Tamil Nadu, Gujarat, Mahārāshtra, Andhra Pradesh, Karnataka, and some regions in Orissa, Kerala and the Himalayas.

Initially a number of single wind turbines were installed and tested in five different states in order to evaluate the possibility of the introduction of large wind power plants to India. The projects included assessments of local wind energy resources, analysis of the local electrical grid system, selection of sites, specification of wind turbines, and drawing up of monitoring and training programmes. In each state a 90 kW wind turbine was installed. The monitoring phase recorded several key parameters and evaluated the local operation and maintenance organisations.

The second phase was concerned with the installation of large wind farms and their integration with the electrical grid. Several aspects were involved:

- Selection of suitable sites with high wind energy resources
- Design of wind farm configurations taking into account the available land
- Preparation of the specifications for the site installation, including the wind turbines, electrical installations and the civil engineering work for the foundations and roads
- Estimation of the economic viability
- Setting up the planning charts for the implementation and for future operation and maintenance activities

The first grid-connected wind farms with individual turbines of 55 kW were installed in India from 1986 in some of the coastal areas. The performance of these wind farms was encouraging and their capacity was doubled. Further site investigations was carried out and in 1990 three large wind farms were commissioned. The farms are sited at Muppandal (4 MW), Kayathar (6 MW) and Lamba (10 MW); all use turbines of 200 kW rated power. A comprehensive training programme was included to ensure the local staff had the technical capability to operate and maintain the plant. By 1993 some 50 MW of wind farms had been installed.

Indigenous capability to manufacture grid-connected wind turbines is being developed in India, and at present almost all parts can be sourced locally. Various financial incentives are offered by the national government and some of the state governments, and they are backed by international aid programmes to encourage private sector investment in wind energy. The incentives include various tax allowances and the ability to 'wheel' power over the public electricity network from a wind farm to a factory some distance away. These forms of support have been spectacularly successful and have resulted in a very rapid increase in the installed capacity of wind turbines in India. By 1996 there were some 600 MW of wind turbines installed in India.

9.6 CHINA

Wind power has been used in China for centuries for grinding grain and water pumping. Because of its huge fossil fuel resources, little attention was been paid to exploiting its wind resource with modern technology until the second half of the twentieth century. Today China is the world's largest manufacturer of wind turbines, but the vast majority are relatively small units for charging batteries and their rotor diameters are less than 4 m. Wind measurements have shown that there are high wind speeds along the eastern and south-eastern coastline, Inner Mongolia, northern Gansu province and Tibet. Wind speeds as high as 11.7 m/s are experienced at Tian Pool in Jilin province.

Unlike India, China has concentrated on standalone and autonomous wind energy systems, which it is able to manufacture totally indigenously. Western parts of China and Inner Mongolia have a good wind resource, but they are sparsely populated and electricity load centres are dispersed, so it is not economic to install a centralised power grid in these areas. Instead standalone wind turbines of up to 1 kW for battery charging and small autonomous wind electric grid systems from 10–100 kW are used to meet the local electricity demand, which is mainly for lighting, radio and TV. About 100 000 of these units are operational in China and in 1990 the country manufactured about 30 000. In contrast there were only six demonstration wind farms in 1992 with a total capacity of 4.3 MW. By 1996 this had increased to 60 MW.

The use of wind-diesel systems is also being investigated. Due to the high transportation costs of fuel oil, the production costs of conventional diesel power plants in remote areas can be high. In order to reduce fuel costs, a pilot wind-diesel power plant was installed in 1988 on Dachen Island about 50 km off China's east coast. The project was supported by the European Commission. Three Danish 55 kW wind turbines and two Chinese 20 kW wind turbines were linked to the local electricity grid using a diesel plant consisting of five generating sets of different capacities totalling 1316 kW. As part of the project, a test station was set up in order to carry out tests on the pilot plant as well as on future wind turbines that may be installed. During implementation a training programme was carried out to enable the Chinese staff to operate the plant and the test station. A major objective of the project was to investigate the possibilities for the local installation of wind–diesel systems in rural areas and especially on isolated islands. Details of the operating experience and an economic assessment of the Dachen installation are given by Wu *et al.* (1990).

9.7 CALIFORNIA

The United States of America had around 16 000 wind turbines installed by 1992 with a capacity of about 1600 MW and generating about 2.9 TWh per year (Ancona *et al.*, 1993). Most of these machines are located in the coastal mountain passes of California, which have the greatest concentration of wind turbines in the world. The winds in this area arise from local land–sea temperature differences and are fairly predictable with few variations from a relatively regular pattern. In the Altamont Pass the windiest month is July and the daily peak is around 1600 hours; variations in the annual mean are quite small. The electricity load curve in California is advantageous for wind power because the demand, mainly for air-conditioning, peaks during the summer months and also in the late afternoon.

The development of grid-connected wind power in the United States occurred rapidly from 1980 to 1985 under the incentives of subsidies for building wind farms through tax credits and by premium payments for the energy produced. The financial incentives may have been more important to the private developers than the technical considerations because the early Californian wind farms had a poor operating record. The stressing of components due to local wind turbulence was not appreciated and many machines suffered fatigue failures from use of inappropriate materials. The wind farms were also a long way from the local grid; this made the long transmission lines prone to faults. Restrictions on the amount of reactive power which could be drawn meant that synchronous generators had to be used, or induction generators with capacitive compensation.

The early problems have been largely overcome and by 1989 one Californian utility reported that on occasion about 8% of its load was supplied by wind farms. Oil-burning power stations were being displaced by wind plant, and the access to hydropower by some of the utilities when the wind was not available was also helpful for the wind plant economics. Loss of load probability analyses (see section 6.3) have shown that wind plants in the Altamont Pass were achieving 9–17% of their rating as capacity credits.

However, in California the proportion of electricity generated by wind power is less than 2% and for the whole of the United States it is much less. An Energy Policy Act passed by the US Congress in 1992 to promote renewable energy sources will provide a tax credit of 1.5 c/kWh over 10 years for new wind plant and give access to transmission lines; it will also give energy production incentive payments to publicly owned plant. These federal financial subsidies should be seen as recognition that the future costs of environmental damage will be rather more than the subsidies being given now to establish clean and renewable energy sources such as wind power. If government encouragement and financial support continues, it is expected that by the year 2010 the installed wind turbine capacity in the United States will grow fourfold to 6300 MW. But recent growth in wind turbine capacity has so far been limited, with only a total of 1650 MW in service in 1996.

9.8 GERMANY

The development of wind energy in Germany has been very rapid over the past few years and by mid–1996 some 1300 MW of capacity had been installed. This growth was stimulated by the so-called 'Electricity Feed Law', which obliges the local distribution utility to purchase wind-generated electricity at a fixed premium. This price varies slightly with location but is a high percentage of the retail price paid by consumers. Thus anyone wishing to install a wind turbine has a guaranteed market for the electricity, so they can develop the project on a firm basis. In addition to this fixed premium, most regional governments also provided some form of direct support for the installation of wind turbines. Most of the early developments were individual wind turbines dispersed along the northern coast of Germany, but more recently there has been greater interest towards wind farms and towards siting turbines in the central and southern regions. Away from the northern coasts mean wind speeds tend to be lower and this has stimulated the development of very high towers (up to 60–70 metres) in order to capture more energy.

SUMMARY OF THE CHAPTER

The big increase in oil prices in the 1970s and the more recent environmental concern about global warming due in part to burning fossil fuels has forced many governments to examine how far renewable energy sources such as wind power can provide for the future energy needs of their countries. Important international collaborative programmes to promote wind energy have been set up by the International Energy Agency and the European Commission of the European Union. These bodies also have an important role in setting internationally agreed standards for manufacturing and operating wind turbine projects. The extensive development of wind power in Denmark – which aims to produce 5% of its electricity from wind generators by the year 2000 – is an example of what a small industrialised country can achieve.

Several developing countries have used wind power for pumping water or for grinding grain over many centuries and are now looking at modern wind turbines for pumping and electricity production. India has embarked on building large wind farms whereas China continues to concentrate on manufacturing many small standalone wind-powered battery chargers for household lighting and television sets. The huge wind farms in the mountain passes in California were an inspiration for wind power development elsewhere but showed that careful siting and good technical design were essential for successful implementation. The quality of life of many people can be improved while safeguarding their environment through the increasing use of the renewable energy technologies such as wind power.

BIBLIOGRAPHY AND REFERENCES

Ancona, D., Loose, R. and Cadogan, J., The United States Wind Energy Program in a Decade of Change, in *Proceedings of the European Community Wind Energy Conferance*, Lübeck-Travemünde, Germany, 8–12 March 1993, pp. 74–77.

Hau, E., Langenbrinck, J. and Palz, W., *WEGA Large Wind Turbines*, Springer-Verlag, Berlin, 1993.

Hunter, R.J and Elliot, G. (eds), *Wind–Diesel Systems*, Cambridge University Press, Cambridge, 1994.

Nielsen, P., Development of Wind Energy in Denmark, in *Proceedings of the European Community Wind Energy Conference*, Lübeck-Travemünde, Germany, 8–12 March 1993, pp. 91–94.

Palz, W., Caratti, G., Diamantaras, K. and Zervos, A., The European Community Wind Energy R&D Programme within JOULE II, in *Proceedings of the European Community Wind Energy Conference*, Lübeck- Travemünde, Germany, 8–12 March 1993, pp. 61–66.

Stevenson, W.G. and Pershagen, B., The Wind Energy Programme of the International Energy Agency, in *Proceedings of the European Community Wind Energy Conference*, Lübeck-Travemünde, Germany, 8–12 March 1993, pp. 78–81.

Walker, J.F., Milborrow, D.J. and Palz, W., *European Wind Power Penetration Studies*, Draft Report under European Commission Contract JOU–00147, 1994.

World Energy Council, *New Renewable Energy Resources*, Kogan Page, London, 1994.

Wu, Y., Qiu, H., Christiansen, H., Sørensen, B. and Nørgård, P., Operation Experience of Wind/Diesel System on Dachen Island in PR China – EEC and China Co-operation Project, in *Proceedings of the European Community Wind Energy Conference.*, Madrid, Spain, 10–14 September 1990, pp. 580–583.

SELF-ASSESSMENT QUESTIONS

1. Name two of the main wind energy programmes of the European Union.

2. Which country has the greatest proportion of its electricity generated by wind power?

3. Which country has the greatest number of wind turbines?

4. Why is the wind regime in California suitable for wind generation of electricity?

5. Contrast the development of wind power in India and China.

6. What policies appear necessary to promote wind power in industrialised countries?

7. Name some measures which are important for the international development of wind power.

Answers

1. JOULE and THERMIE

2. Denmark

3. China

4. The annual and daily wind variation broadly follows the electricity load, which is mainly for air-conditioning: high in summer and peaking in the afternoon. It is also fairly predictable.

5. India has seen the need for the phased development of grid-connected wind farms whereas China has promoted small standalone wind generators. Note that China is a much larger and geographically more diverse country than India.

6. Environmental considerations, non-reliance on imported energy sources such as oil, financial incentives and subsidies; see also Chapter 5.

7. Developing internationally agreed standards for wind measurement, turbine testing, etc. (IEA); quality control and certification of manufacture and installation of wind power plant (Denmark); dissemination of research results and operating experience through international meetings and journals.

Index